D1139435

LEEDS COLLEGE OF BUILDING
T10528

QUALITY THROUGH DESIGN

QUALITY THROUGH DESIGN

The key to successful product delivery

John Fox

McGRAW-HILL BOOK COMPANY

London · New York · St Louis · San Francisco · Auckland · Bogotá
Caracas · Lisbon · Madrid · Mexico · Milan · Montreal
New Delhi · Panama · Paris · San Juan · São Paulo · Singapore
Sydney · Tokyo · Toronto

Published by
McGRAW-HILL Book Company Europe
Shoppenhangers Road, Maidenhead, Berkshire, SL6 2QL, England
Telephone 0628 23432
Fax 0628 770224

British Library Cataloguing in Publication Data

Fox, John
Quality Through Design: Key to Successful
Product Delivery
I. Title
658.5

ISBN 0–07–707781–4

Library of Congress Cataloging-in-Publication Data

Fox, John
Quality through design: the key to successful product delivery/John Fox.
p. cm.
Includes bibliographical references and index.
ISBN 0–07–707781–4
1. Design, Industrial—Management. 2. Manufacturing processes—Quality control. 3. New products—Management. 4. Production planning. I. Title.
TS171.4.F693 1993
658.5′752—dc20 92–43602
CIP

1234 CUP 9543

Typeset by MFK Typesetting Ltd, Hitchin

and printed and bound in Great Britain at the University Press, Cambridge

CONTENTS

PREFACE

Many industries in the field of product design are becoming increasingly worried today about their competitiveness. They are right to be worried about this and the fact that designing products for today's world needs a new and innovative approach. No longer can we survive on the old traditional methods that served us so well in the middle part of this century. One has only to look at the industries that have been eroded in the West in the last few decades to appreciate the trend. Key industries, like shipbuilding, motor cars, electronics and business machines, have all come under threat and many have declined altogether.

What happened? Industries that are struggling for survival today will tell you that our inability to bring products to market in a timely way is the biggest single factor that threatens us in the face of the worldwide competition coming mainly from Japan and the emerging nations. However, other factors, like cost and reliability, are also high on the list of shortfalls in our present methods of product innovation.

We have to find ways in which we can both reduce our time-to-market schedules and at the same time satisfy customers fully in terms of performance and value for money. If we do not deal with these aspects our competitors will move in and beat us to the winning post.

Why have we become so uncompetitive in product design? After all, we once led the world in technology and innovation. The answer may be that the methods by which we design and develop products have not changed fundamentally since we were in this leadership position. We have relied almost exclusively on designers to give us the quality in our products and have neglected to change the ways of improving our overall design efficiency in a changing world. For instance, we still rely on testing, testing and more testing to establish functional capability of the product. It is of course true that we have made management and production quality control changes in our industries and this has certainly changed the way in which we check and monitor progress throughout the development of a new product, but we continue to rely on the old traditional methods of design with the result that we spend far too much time fixing problems that emanate from a poor original design.

This book takes a new look at what design really is and proposes some new methods which actually make changes in the way designers do their work. Primarily it addresses function in a way that has rarely been done before and shows how we can deal with functional performance in the design without relying on testing alone to confirm it.

Early in the century the design engineer led the whole innovation process from concept to delivery of the product and this meant that a single mind oversaw the whole picture. As mechanical engineering expanded its horizons, the engineer needed drafting personnel to carry out the laborious work of converting ideas into drawings that could be used to manufacture the product. This group of drafting experts evolved into the traditional drawing office team and a new culture emerged where the engineer became detached from the process and relied on the designer to use experience, develop standards and take over the responsibility of bringing quality to the design of the product. Thus developed the two elements of the mechanical engineering process, the engineer and the designer.

Educational trends did not consider the skills of the designer to be important to the engineer, but rather considered that more basic science and analytical techniques should be included in the engineering curriculum. As a result, the science of engineering design has been neglected both by industry and academia, leaving a gaping hole in the sequence of innovation, design and manufacture. This is manifested in the abundance of poor quality goods and the apparent inability of the Western world to compete effectively with Japan in terms of cost, quality and time-to-market. Western industry has attempted to redress the situation by concentrating largely on looking for solutions 'downstream' in the manufacturing process, robotics, quality control and improvements to manufacturing process control. *However, you cannot put manufacturing quality into a design that does not have quality designed in*.

There is an urgent need to bring understanding, science and discipline to the process. This book does that. It does not treat the process with mathematical precision, but there are many books covering the great variety of analytical techniques available. It defines the process of design, treating it in its broadest sense from concept to manufacture and maintenance. It identifies some of the techniques and disciplines required to deliver quality products and establishes processes that help to monitor quality throughout the cycle.

The book is intended to complement the analytical and scientific skills already established in today's modern engineer. It seeks to eliminate the trend towards *design-by-iteration* and encourages *design-for-latitude* which is essential if we are to match the efficiency of the Japanese methods. Additionally, it addresses the role of management in the process to ensure proper monitoring and decision making as the design progresses.

The book centres around the concept of managing the critical functional parameters of a design. This concept enables effective processing of conceptual ideas while providing the vehicle to continually control the design through the whole process. It also matches well with other well-known techniques for focusing on quality, such as Taguchi, quality function deployment, etc.

The aim of this book is to provide structured guidelines to the designer, the engineer and the product manager working as a team to develop mechanical engineering products. Throughout the book, the titles designer, design engineer and engineer may be used interchangeably. In general terms, of course, the designer is recognized as the person who puts the lines on the paper to create the drawings. Also, in general terms the engineer is regarded as the expert on how and why things work. In reality, however, there is a massive overlap between these two roles, and the extent of this overlap often depends on the particular bent and choice of the individual. Most of the experience seen in design offices has been generated by passing on knowledge and techniques that have been developed through experience over the years. This book aims to provide a set of guidelines and disciplines that can provide a framework and structure to the mysterious and complex process of design. It will take the reader through the questions of what we mean by design and what is the thought pattern that goes on in the best designer's head. It will try to broaden the concept of design from the popular idea that it is just a set of drawings or sketches to the wider sequence of where it starts and where it finishes. It will also try to lay out the fundamental steps of such a process and go on to analyse each of the steps, giving a coherent and rational structure to the events.

The last chapter in the book is devoted to managing the design process and deals with the subject particularly from the designer's viewpoint. In other parts there is reference to management processes but only when this has a direct bearing on helping the designer or the design team to do their job more effectively. The design process is generally thought of as a team effort and in this respect the popular view is right, as designing involves many linkages between people, events and functions. Thinking of it as a team effort, however, is not enough. It is the type of team effort that must be understood. It has been said that designing is like a Rugby match where the members of the team continually interact until the game is over. It is not like a relay race where the responsibility is handed from one member to the next until the race is over. Many organizations do operate in this way, the design office preparing the drawings for manufacturing to make, the only relationship between the two being at the hand-over of the drawings themselves.

Whoever reads and follows the recommendations of this book will have to make significant changes in their product development processes. Some of these changes will require the training or retraining of their personnel. On a broader scale it will take radical changes in our universities and our industries to bring about these proposals. In the end there is no alternative if we are to meet the ever-increasing threat from a competition that never stops improving its effectiveness.

Only through design will there be quality.

Note

The author of this book now provides a consulting service supporting improvements in design quality. He can be contacted on (0438) 740946.

ACKNOWLEDGEMENTS

A number of the ideas represented in this book have been contributed by various colleagues involved in product design. In particular, I should like to acknowledge the following people for their contributions:

Ron Feldeisen for his ideas on the design process
Bruce Parks for the architectural process
Maurice Holmes for definitions and ideas on technology readiness
John Strutt for his design monitoring process
Hiro Matsubara for his unique knowledge on the differences between the Japanese and Western cultures

There are many other people in Xerox Corporation and Rank Xerox Ltd that I should also like to thank whose ideas I have absorbed over the years and which are included here.

Other material has been kindly supplied by American Supplier International, Ltd (the quality lever and quality function deployment diagrams) and by Rexel Ltd.

I should also like to thank the people who have helped, encouraged and supported me in writing this book, namely my wife Val, Roger Ryder and Malcolm Thayer.

John Fox

To Val, Sarah and Matthew

CHAPTER

ONE

DESIGN

1.1 THE NATURE OF DESIGN

One of the approaches to delivering better products is to try to understand the complexities of design and search for ways in which the process can be improved. The start to this exercise should perhaps be to find a good definition of what design really is. Good definitions are hard to find. The Oxford Dictionary describes design as 'conceiving a mental plan for; making a preliminary sketch, picture'. Others who are closer to the discipline have described it in more detailed terms. Hollins and Pugh (1990) describe 'total design' as 'a multidisciplinary iterative process that takes an idea or market need forward into a successful product'. They note that design does not end with production but with product disposal. Middendorf (1990) defines design as 'an iterative decision-making activity whereby scientific and technological information is used to produce a system, device or process which is different in some degree from what the designer knows to have been done before and which is meant to meet human needs'. Bebb (1988) describes design as 'the set of processes that translates customer requirements into manufacturable outcomes'.

Some points that are certainly true about design are:

It is creative
It is a multidisciplined process
It seems to need to be iterative
It is evolutionary
It serves human needs.

Quality is also in need of clear definition. Some people argue that quality is that attribute held by such a product as a Rolls-Royce motor car. Although this is accepted by most people as being a quality product the definition needs to address quality in much broader terms. The most acceptable definition which seems to satisfy the needs of the product designer is that quality is simply 'meeting the customer's requirements'. A Rolls-Royce no doubt satisfies this definition but likewise any car that meets the expectations of the customer in all respects also satisfies the definition and one of the requirements may be low cost! So perhaps a Ford Escort or Citroen 2CV could be termed a quality car if it meets the expectations of its customers. These expectations would naturally include reliability, running costs, etc.

Combining both definitions of design and quality we can infer that design quality is 'The processes and activities that need to be carried out to enable the manufacture of a product that fully meets customer requirements'.

There is acceptance today that the concept of a design team handing over a design to manufacturing is not the way to do it. Gone are the days when the product designer's work was over with the delivery of a set of drawings for the manufacturing manager to make. This 'hand-over' practice was employed not so long ago and manufacturing would be quite justifiably appalled to be expected to produce a design that had no consideration with respect to the manufacturing process, product quantities, tooling, etc.

Concurrent or simultaneous engineering is now widely accepted as the best and only way to go. This embodies communications and interactions with a wide variety of functions: marketing, manufacturing, business planning, finance and servicing. All these entities play a role in the activities of product delivery and have to work together and communicate effectively to bring quality to the process.

From the designer's point of view it is important to distinguish between the need to deliver quality drawings and the need to develop the correct information for these drawings. Making a complete and thorough input into the processes that ultimately deliver the drawings themselves is paramount to design quality. As an example consider a designer designing for a simple drive system consisting of an electric motor driving a pulley under a torsional load via a toothed belt. Quite often this will be done quite intuitively by a designer who will do no more than put ideas down on paper as a set of drawings. The outcome is normally satisfactory because the designer quite likely has sufficient experience and inherent knowledge of such systems. However, without any understanding of output load, load capability of the belt, input power of the motor, whether intuitive or not, the design will be prone to failure. In any case, unless the engineering analysis of the system is completed at some minimum level, the system design can at best only be overdesigned. Overdesign can be just as great an enemy to design quality as inadequate design, since it inevitably affects cost and other important aspects of the product such as schedule and space, all of which may be important to the customer and will therefore make the final product less competitive. Often the design quality is entirely dependent on the quality of the designers on the team because the project relies on their experience and knowledge. I have heard chief engineers say, when discussing how to get a good design completed quickly, 'Give me my choice of designer and I'll give you a quality design.' While this is one way of doing it, it is essential to put some structure around the process in order to maximize the performance of every member of the team. After all, it would be an extreme luxury always to be able to run a design team with only the very best designers. Additionally, the intuitive approach leads to a design that is rarely optimized. Such a process leads to a style that uses the iteration of hardware to solve design problems. This is a path which, as we will see later, is expensive both in terms of money and schedule. Even though optimization may have been regarded as a luxury in the past, today, with fierce competition from Japan and Europe, optimization is essential to get that extra quality which will enable modern industries to gain that competitive advantage vital to their survival.

A study of people's perception of what design is revealed some interesting insights. A group of engineering personnel, spanning disciplines including laboratory technicians, design engineers, managers and designers, was asked to state in a few words how they would describe the design process as they saw it. The results were analysed for references to a number of different attributes considered key features of the design process. Each description submitted was analysed for reference to:

- Generation of concepts
- Understanding operation or function
- Translation of information for manufacturing
- Conversion into hardware

The results showed that most people in the group felt that generation of concepts, translation for manufacturing and conversion of ideas into hardware feature heavily in their concept of the design process. There was, however, a markedly low vote for understanding in terms of the function of the design. Although not statistically valid, this study does highlight the tendency for people to think of design only as a creation and implementation process, neglecting the vital intermediate step of functional understanding and optimization. This has led to a process whose style is to create, try-it-out, fix-it, or more simply a process of iteration.

Iteration as a method for improving quality and reliability has been shown quite conclusively to be time consuming, costly, inefficient and ineffective. It has developed from the gradual acceptance into the engineering profession of unqualified and untrained people who have been allowed to practise the profession. There have been many occasions when I, as a development engineer, saw untrained people 'have a go' at resolving a problem by trying something that looked as if it might do the trick. It often appears to be the easy way out of a problem to the untrained eye. No doubt in the development of a simple product, iteration can be as effective as any other method of improving the design. However, once a certain, fairly low level of sophistication or complexity is introduced to a product or system, iteration can no longer be employed as the best method of improvement. For example, a customized chip for an integrated circuit has to be correct before the mask is made, since the mask may cost as much as £100 000, the chip costing only pence to manufacture. Any process of iteration in this case to get the design of the mask right would be absolutely prohibitive.

A further look at what the design process involves helps us to appreciate this point better. In seeking a high quality design we are looking for ways of achieving what the customer wants in all its aspects. Let us consider a consumer product that we are about to design. Some will argue that all the customer wants is a product that functions effectively and reliably. Very often, one hears this comment made by a consumer in respect of cars, washing machines, etc. What is implied here, but often unsaid, is that for instance it must be done at an acceptable cost. From the point of view of the business situation, it must also be completed and on the market prior to competitive products in order to gain an adequate market share. It must look attractive, not be too noisy, not pollute the atmosphere and it must be able to be manufactured and serviced. Each one of these design requirements has an effect on each of the others. For example, we can make the product more serviceable by adding, say, quick-release features, but these will cost more and may even detract from the reliability of the product. A cheaper motor or fan may satisfy the requirements of the function but may be more noisy, and so on.

So one could regard the process of a high quality design as one that provides the best balance between all the requirements of the customer. The achievement of this balance is a highly complex process of communication between different disciplines and an equally complex process of managing the relationships and interactions of these interfaces. As a design engineering manager I had an amusing vision of my job being analogous to the entertainer who starts a series of plates spinning on sticks. The entertainer starts with just one or two of them spinning, but has to leave these to start others alongside. Soon the initial plates need to be rejuvenated and this requires the artiste to go back and bring them back up to speed. The object of the exercise is to get all the plates spinning at the same time. Soon it becomes clear that the artiste must prioritize the time between starting new plates or maintaining those already in progress. Design is a bit like that in that as a design manager one has to give attention to each aspect of the design process and no one element can be left alone for long without needing some attention, either to solve a problem or to deal with some interaction that has occurred.

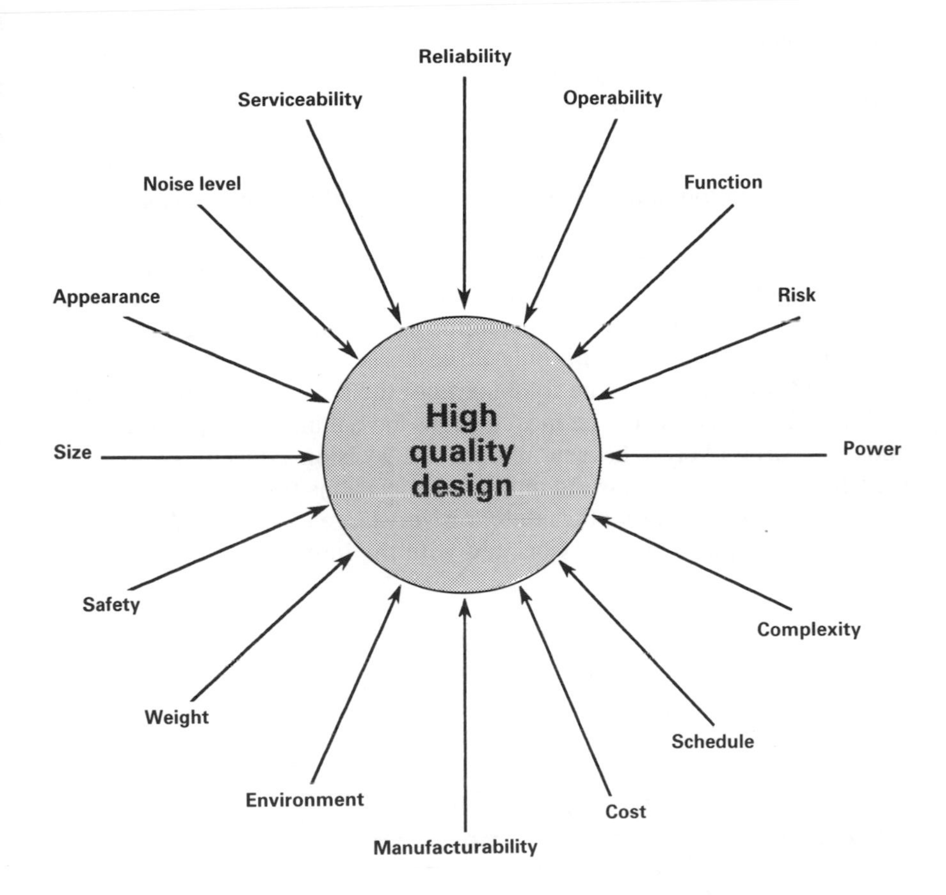

Figure 1.1 The complexity of design.

Figure 1.1 shows the typical set of interactive elements present in any design exercise. The individual designer's job is to design a particular part of the system trying to satisfy all the demands from these forces and with their interactions in mind. Every element of the design, whether it is a part, a subsystem assembly or the complete system, has to comply with the demands of each of these attributes. The number of decisions in this process is enormous. Is it any wonder that we find ourselves as designers in a process that makes a 'best attempt' at this and then checks it out by testing to see if it works? The effectiveness of the process must be improved if we are to reduce schedules and costs.

The number of decisions made during the normal process of design is enormous. To illustrate this, consider the simple case of making an early decision in the designing of a car. The car could have front or rear wheel drive and front or rear mounted engine. The combinations are:

Front wheel drive–front engine
Front wheel drive–rear engine
Rear wheel drive–front engine
Rear wheel drive–rear engine

Hence a total of four alternatives are available in this early decision process for a car design.

In more general terms, the total number of decisions made by a design team during the process of the development programme must be very large indeed. If we take only three subsystems and assume that each subsystem contains assemblies each with only three parts and has only one decision associated with each part, the complexity of the decision matrix is depicted by Fig. 1.2. In fact, of course, most designs have many more parts than this and each part is likely to have more than one decision associated with it. It is therefore difficult to assess the number of decisions that need to be made, except of course to say that the number is very large indeed. Added to this is a mass of decisions that are associated with the interactions between parts

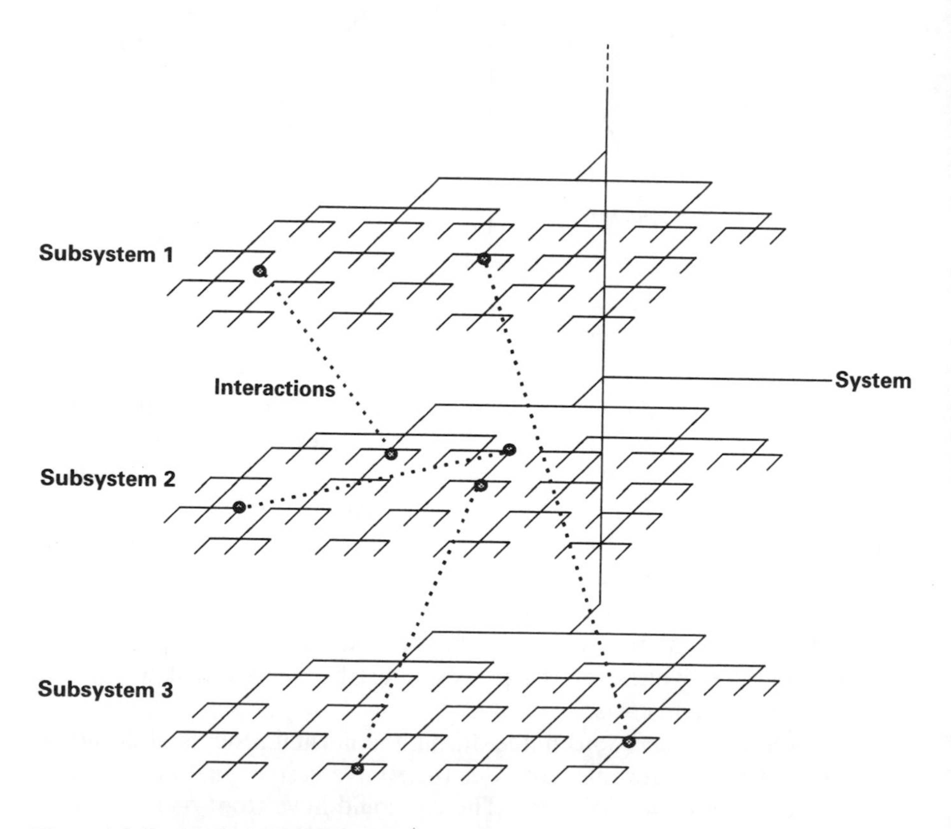

Figure 1.2 Design decision matrix.

(only four are shown in Fig. 1.2) and these interactions add further to the complexity of design decisions.

It is not surprising that there are many opportunities for error in the design process. This is confirmation that the process needs to be dealt with in a structured manner so that the complexities can be dealt with effectively and with minimum error.

1.2 THE COMPOSITION OF DESIGN

There are two main problems in dealing with the overall concept of design. The first is that design is a very complex process and, secondly, confusion is often caused because design means different things to different people. In dealing with the complexity question it is obviously important to break it down into handleable elements which can be understood and analysed. In terms of exactly what we mean by design it should be made clear from the outset that design is a process and not the outcome of a process such as a set of drawings or the product itself.

For the purpose of this book and for the general engineering graduate engaged on the design of a product, it is necessary firstly to understand the elements of design and also the various stages of the design process. It can be thought of as slicing design into elements in two different dimensions, one being the physical plane and the other the temporal plane.

The word design has been broadened so much that it can mean anything from a new example of high fashion to a piece of equipment that performs a complex function. The increasing use of the word 'designer' to give a product market appeal and credibility presumably implies that a well-known designer has spent some time creating the product specially for the customer. It is applied mostly to the artistic end of the spectrum of product design. By contrast a device such as a robot, used to assemble components automatically, will have been designed almost entirely with function in mind, little concern has been applied to the appearance of the device. In a way this spectrum of design definition embodies all the elements that are found in engineering design. Every product that is designed contains three elements of design (see Fig. 1.3):

1. *Form.* This is the aesthetic or artistic end of the design spectrum. It is what the public today usually means by design. The shape and appearance of the product is quite often regarded as 'the design' in today's parlance.
2. *Fit.* This is the part of design that deals with how things go together, how they integrate to form the whole system. It is addressed well in engineering by traditional limits and fits laid down by years of experience and analysis.
3. *Function.* This aspect of design is the one that most often dissatisfies

customers. It is the aspect of design that causes failures or poor performance of the product and is perceived by customers as not meeting their requirements. Interestingly, if function is designed well and continues to perform well, the customer will not necessarily be fully satisfied, since it may be aspects of form or fit that are perceived as inadequate. Perhaps if function were the most important aspect of a product for the customer, we would keep our cars longer than we do at present.

In considering these three elements of design and in questioning how we can improve the quality of these to enhance the quality of the overall design, it is important to note that if we want to improve quality we must be able to measure each of the elements in order to assess if and when quality has been improved. So how can we measure each of the three?

Form can be measured only by judgement. Form, like beauty, is in the eye of the beholder and as such is dependent upon individual taste. Something that is pleasing to one person may not be as pleasing to the next. Quite often the only way in which form can be measured is by obtaining consensus

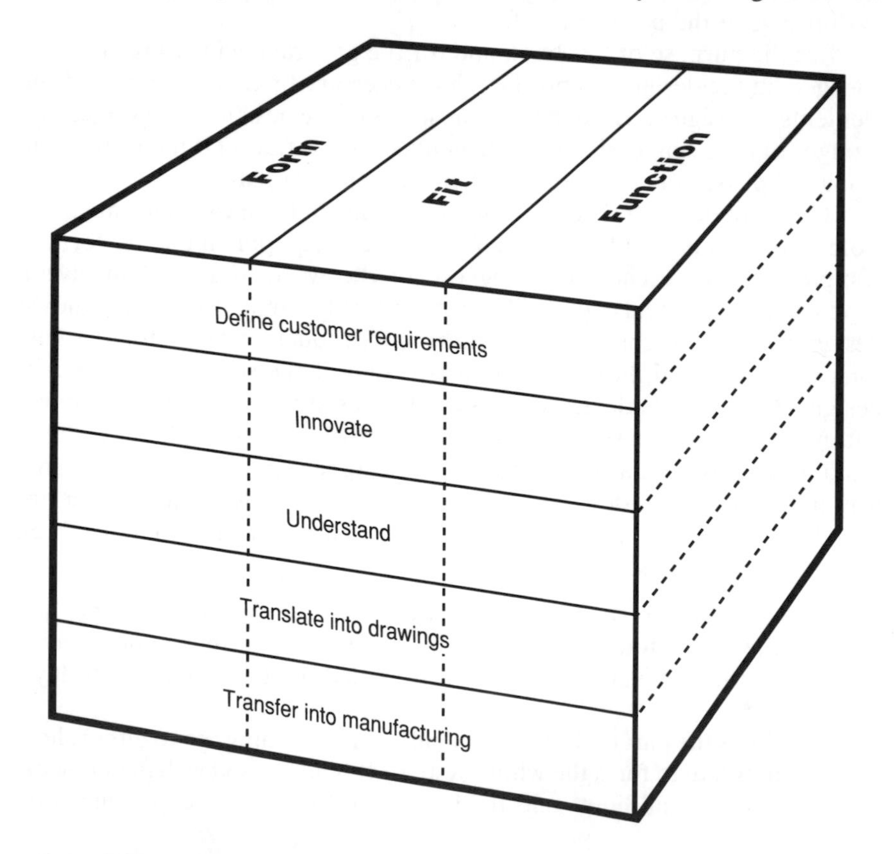

Figure 1.3 Elements of the design process.

or collective opinion. Experts in industrial design may be able to predict quite accurately whether one particular aesthetic design is more appealing to a particular customer than another. Indeed, sometimes the entrepreneurial industrial designer will deliberately attempt to set a new fashion by leading with something outrageous. But these are few and far between. More generally form is measured and assessed by testing opinion and drawing conclusions from that.

Fit is much easier to measure. In fact, engineers have addressed their attention to measuring fit since the dawn of engineering. Handbooks on mechanical engineering, which list recommendations for fits and standard components like bearings, have well-established data to ensure that their fit is optimum for proper operation and life. Checking fit is the primary tool of the inspection facility or quality control in the production environment, and makes the basic assumption that the design was correct in the first place. There are simple tools and highly complex tools for measuring fit and any aspect can be fully quantified with no opinion needed to confirm.

Function is not such an easy area of design to measure. Traditionally the process for establishing correct functionality has been to test. The actual measurement of effective function during this testing is quite often to look for failures. Alternatively, the test result can be measured against a laid down specification, setting limits of acceptable operation as thresholds for unacceptable performance. As we will see later, one of the problems with testing for good functional operation is that one has to build hardware and incur heavy costs to do so. Of course many industries now use simulation as an alternative to testing and, prior to that, analysis, both of which are infinitely less costly and should be pursued more and more.

Therefore, if we are to improve design we must improve it in one or all of these three areas, which means that we must measure our success in each of the three and search for ways in which to achieve a better result.

The other way to slice design is in the temporal plane. In other words, what steps do we go through to proceed from start to finish of a design? Again, if we are searching for ways to make improvements we must look at each of the phases of design, be able to measure them and try to find ways in which a more effective result will be produced. Each of the phases is a candidate for improvement.

Figure 1.3 shows design as a cube which is made up vertically of the elements form, fit and function. In the horizontal plane the design cube is divided into five temporal elements of design. Let us now explore these five phases of design activity with a view to understanding how each could be improved:

1. *Define the requirements of the customer.* This may be one of the most neglected elements of the design itself. Neglect of this will, most obviously, not deliver a product that satisfies the customer's needs. Indeed, it may deliver a product that the customer does not want. Neglect of this

does more. It inhibits the progress of the design team, often causing them to make wrong assumptions and then, late in the process, bring in changes to correct the situation. If the design team does not know in precise details what the customer wants then the design that they ultimately deliver will at best be late, is likely to be inadequate and worst of all will not be what the customer wants and therefore a business failure.

To prevent this situation arising it is essential that the design does not proceed until all the detailed questions about requirements have been answered in full. It is not permissible to answer a requirements question 'at a later date'. This will inevitably cause major disruption to the design itself. Customer's requirements can be established by market research, looking at what the competition is doing and by getting feedback from the field on existing products.

2. *Generate ideas by innovation.* This is one area of design that most people recognize as a key element. As such it is quite often done well. Choosing the best way to innovate can help to improve results. Later in this book I will talk more extensively about processes that help to do this and how to make decisions on the best alternatives to adopt.
3. *Understand the function.* Once a method has been chosen to implement the design, whether innovative or not, it is important to go through an effective understanding of what drives the function, what is the output from the function and what failure modes might be expected. Again this is dealt with in great detail later in the book. If this element of design is not fully carried out there is only one alternative to ensuring that performance criteria are met, and that is by testing hardware.

   Unfortunately, the process of testing hardware means that if the result is not fully successful the design has to go into an iterative style of fixing problems in the hardware and retesting. This is not an efficient way of getting to a successful design, even though it is a process adopted by many. Understanding the function correctly in the first place is a major benefit to the subsequent steps in the design as it eliminates repeated iterations of hardware to implement changes and can streamline the solution to any problem that does arise subsequently.
4. *Translate the design into drawings or a design file.* In theory, drawings are irrelevant to the design process. If one could take the innovative ideas, understand their functionality and transfer them into a production environment, drawings would be unnecessary. Unfortunately, except for some modern techniques like SICAM (supplier integrated computer aided manufacturing), this is not possible and so drawings in some form or other have to be prepared to bridge this gap. However, the production of drawings is simply a translation of the ideas of the innovative designer into something that can be read effectively by others. It is *not* the design itself and moreover the quality of the design *cannot* be assessed by the quality of the drawings. Many people think that the process of producing

the drawings is the design and that it is the only real element of the design. Of course this is not the case. Design is good or bad according to what goes into the drawings, not the drawings themselves. Nevertheless, if drawings are needed (and they are in most cases), they must be of high quality and free from error. Later in this book some methods for ensuring the quality of the drawings are described.

5. *Transfer the information into manufacturing.* Strictly speaking, transfer of knowledge into manufacturing does not occur sequentially at the end of the process. We should guard against the idea that a set of drawings is 'handed over' to production once the design has been completed. No doubt this used to occur in some industries years ago, but in today's modern competitive world it will not work.

   The capability to manufacture the product and the parts must be designed in throughout the whole process. Ideally a manufacturing expert should be involved in the design from the outset. (Ideally the manufacturing expert should be the designer!) Design for manufacture and design for assembly are integral parts of the design process and should not be neglected. The development of the design from the early time with parts made by hand or from soft tools to the later time when a high percentage of hard tooling is used is a difficult transition, demanding the talents of both disciplines of engineering.

1.3 PERCEPTIONS

Art versus Practice

Following the idea of design as a combination of form, fit and function leads to the concept that the design process must also have an element of each of these in it. If this is so, then the design process must itself contain a spectrum of activities ranging from the aesthetic to the practical or operational.

Design is essentially a practice of creation. It requires the practitioner to deliver new and innovative solutions to express function. It also requires that form be created to bring an element of beauty to the resulting product and it demands the skill of depicting form, fit and function by drawing. Modern-day drawing of course occurs both in the conventional way using lines on paper and also with recent computer technology by organizing electronic lines on a cathode ray tube to create a design file.

As we have said, design is a creative activity that is present in men and women of many different leanings. In a modern-day design team this range of attributes comes together in a variety of professionals, each trained in a different skill. Every design (whether it is an engineering design or some other kind) requires varying amounts of creativity, ranging from the aesthetic to the practical. For instance, the design of fashion clothes requires a high proportion of aesthetic creativity, but much less of the practical or functional creativity. On the other hand, a robot, which can rarely be said to be designed with artistic flair, has a very high proportion of functionality involved.

In engineering design there is a combination of both these aspects of the design skills, the proportions depending very much on whether the product is an item with a requirement for artistic appeal or more functional in its basis. If we look at the skills that form the generic team of engineering designers it is interesting to consider where each member fits within this spectrum.

As we move towards the practical side of the process, design needs the knowledge of materials that are both traditional and modern. Indeed, if we are to be competitive, modern plastics and structures are a key area of knowledge for today's designer. It requires the ability to analyse these materials and the training to understand the physics that applies when dealing with them.

Product design is a business and requires the knowledge and skills to determine cost and the ability to take decisions associated with cost. Finally, the process of manufacturing parts and assembling them is a skill that must be incorporated into the design process rather than added on at the end.

All these skills come and go as the design progresses. If we look at how these skills are spread across the members of a typical design team it helps to understand how each of the members views the design activity and where the emphasis or preferences might lie. Figure 1.4 shows a matrix of the skills that form this generic design team against the elements of the knowledge base for a typical design. The primary skills for each contributor are shown shaded, while those of a secondary nature are shown unshaded.

The artist's role is to bring aesthetic excellence to the design of the product. The aim is to appeal to the customer with beauty and an appearance that does not intrude excessively into the customer's world. The artist needs to be skilled in creation, form and drawing, and to a lesser extent innovation.

The industrial designer adds practical aesthetics to design, usually in the way of form, and will become influenced by the ergonomic and human factor aspects of the design. These skills move towards the more practical end of the scale with emphasis on the knowledge of materials, particularly as used to enhance appearance and form.

The concept designer is the central figure in the design activity and must have knowledge of all aspects of the design from the materials that enable the shaping of the form required by the industrial designer to the costing of each part to be manufactured. A sensitive path must be trod between all the aspects of the design to deliver a balance that is acceptable for a successful product. The skills of the concept designer are broad and understanding of all aspects is extensive.

Art ⟶ Practice

	Artist	Industrial designer	Concept designer	Product designer	Detailer
Creation	●	●	●	○	
Form	●	●	●	○	
Innovation	○	●	●	●	
Drawing	●	●	●	●	●
Materials	○	●	●	●	○
Analysis		○	●	●	○
Physics		○	●	●	
Costing		○	●	●	
Manufacturing			●	●	○

Figure 1.4 The skills of design.

The product designer is less involved with creativity and form and more with practical considerations such as cost and manufacturing.

The attention of the detailer is focused on the construction of the drawings and the documentation which gives the design its final definition, recording every element of the process of manufacturing as an information package for the future.

Management Process versus Design Processes versus Tools

As competition increases and customers become more and more demanding in their requirements, companies are searching for ways of improving quality and reducing schedules. Almost every technical journal that one opens contains some new proposal to improve the way that design is carried out. These new methods address various aspects of the design process and often there is confusion in terms of how they contribute to improvement.

It is essential to separate out the various elements that affect the way the design team goes about its business of designing. In the eighties it was fashionable to look at the processes in manufacturing in an attempt to improve quality and performance, and many good practices came out of that era. Typical was the concept of on-line quality control which ensured that the quality monitoring of the manufacturing process was carried out in a way that gave continuous control of the quality of the output. Robotics and design for assembly was another step forward which enabled improvements to be made in the consistency and final quality of the assembled product.

Alongside this came many attempts to change the way in which projects were managed. More structure was introduced into the management process with more monitoring and redirection where necessary. Processes for checking and measuring the progress of design in association with manufacturing became popular and it was generally understood and established that design for manufacturability was a key to good product quality. In addition to these the product development industry was faced with a massive growth in the introduction of computer technologies which changed the face of communications and data management forever. The extensive use of computer-aided design techniques, computer-aided analysis and engineering grew rapidly, to the extent that design managers had great difficulty in deciding which system was the best one for them to use. All these various pressures and influences were brought to bear on the traditional design team. In parallel with this, competition became increasingly fierce and caused companies to look at ways in which they could meet Japanese benchmarks in terms of staffing levels and other resources. This trend goes on as we strive to at least keep up or at best take the lead in our competitive environment.

In pressing forward with these new and exciting ideas, we must get the perspective clear when considering new ways of developing products. There are three distinctly separate groups of changes that can be introduced into the process, and they are often discussed as if they all fit the same category:

1. *Management processes.* These are processes that affect the way the team is managed, such as motivation, resource allocation and monitoring progress. They can be used to change the outcome of the design direction or emphasis, but they are often introduced as if they could change the design itself. Essentially they do not do anything to change the way in which the design is done. They may introduce a new process for issuing a drawing which ensures that it meets certain standards. They may provide a means of deciding whether a design is adequately matured to meet the expectations of the programme. However, the outcome of the design depends on how effectively the team converts the ideas into the final product. Although such a management system must be effective, it has to be limited in how much it can change the outcome of the design.
2. *The design process.* Changes under this category are changes that tell the designer or the team how to do their job in a different way. If the management process is a map to get from A to B, then the design process is a lesson in driving on the road that the map defines. Unless we tell the designer to draw lines on the drawing in a different way, we will not significantly influence the change in the outcome of that design. This re-emphasizes the central theme of this book. The book is intended to make proposals as to how the designer and the team should change the

way in which they do business in order to get closer to a higher quality product.

3. *Tools and methods.* This is the area most often confused with changes in the design process. Many companies think that the introduction of a new tool or method will dramatically change their efficiency. If it takes 30 hours to produce a drawing using a drawing board, they may think that this can be reduced by a significant percentage using computer-aided design (CAD). Or if they can use a stress analysis programme, this will eliminate errors in structural failure that have been experienced in the past. On the contrary, the introduction of a new tool is most likely to reduce the efficiency initially while the training and learning processes are in progress and before any benefits are seen to take effect. If a company changes from manual drawing to CAD, it is likely that the rate and quality of drawing production will reduce at first. Added to this is the idea that any new tool or method will require new support functions to enable the implementation of the new scheme. Tools and methods can be used as part of an overall programme to help support improvements in quality, etc., but they are not the panacea for all ills and must be introduced as part of a broader programme for improving the effectiveness and productivity of the team.

Enablers for Design Quality

When considering what it is that goes to make design quality, we have to look at the subject from all possible angles. During a study that I was involved in to try to enhance design quality, an analysis was done of the various aspects of the process that actually enable the design, that is what needs to be in place for the process of product design to go ahead unhindered. The result of the study gave no surprises. A number of key factors have to be in place and these are obvious to anyone looking at the results, and yet experienced designers will ask how many times these 'obvious' elements are not in place.

Design quality can be achieved if the following enablers are established at the onset of the design exercise (Fig. 1.5):

1. *Design inputs.* The information that the design team needs to get on with the job and not be hindered for lack of one small piece of information. As we have seen elsewhere in this book, there is an unquestionable need to establish at an early stage the goals and customer requirements of the product to be designed in the most unequivocal and complete terms. Goals are defined by the marketeers, but the designers need more than goals. They need to be able to understand the specifications that will enable the marketeers' goals to be met. They need unambiguous details of what the product inputs are, what the outputs are and what are all the constraints or boundaries to these. These details must be worked out

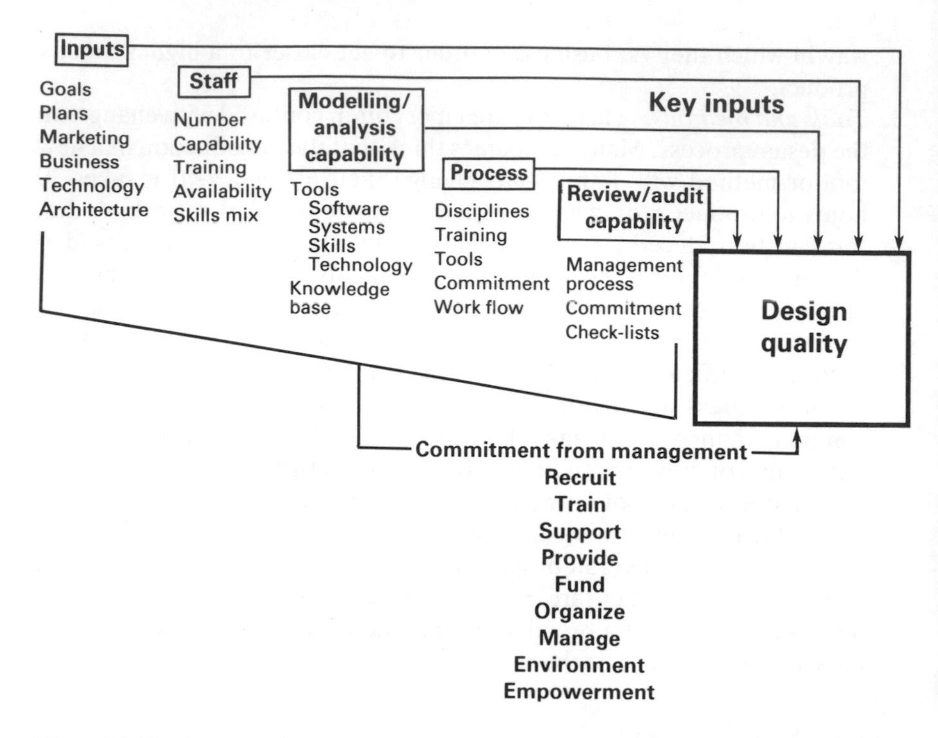

Figure 1.5 Key inputs to design quality.

using marketing information on customer needs, together with technical information on capability and constraints due to cost or schedule. The basic technology set must be clearly defined to the design team, as must the architecture of the product. It is often helpful to the designer to provide marketing details describing why the product is needed. Details such as which products are being targeted and the particular features which need to be competitive are extremely useful to support goals and customer requirements. All this, supported by a viable business case and a workable plan, will provide the necessary input material to enable the design team to set out on their task unencumbered.

2. *Staff*. The human resource in terms of total power and expertise to tackle the job. Assuming the team has the necessary input material, it will not proceed very far along the line unless it has the proper resources to carry out the work. This does not mean just the right number of people to do the job, but the right number with the right skills and training. The team must also have the correct balance of skills and must be available to work as a team. All this sounds obvious, but bringing this all together fully at the outset is a task rarely accomplished.
3. *Modelling and analysis capability*. The necessary academic expertise and tools to support the level of technology being used. The need to model

and analyse various parts of the design varies enormously with different products and levels of technology. Therefore the first thing to decide is just what level of analysis will be required and then to establish the capability through providing the right knowledge base and the right tools to implement the processes. Tools could range from, literally, a piece of equipment to emulate some design aspect to a computer program, such as Monte Carlo or Nastran, to provide the power of many calculations supporting the design aspects.

4. *Processes*. Those disciplines and procedures that enable control of the design to take place. These are the disciplines that we ask our design team members to exercise to ensure that the design can be monitored and kept under control. For instance, there are many engineering processes and disciplines that must be employed, such as configuration control to ensure that the bill of materials and the process of change is always under control and understood. Another example is the way in which the cost of the product is recorded and monitored as it evolves through the phases of the design. These and many more processes must be in place and defined prior to the start of the programme. If they are not, inevitably some aspect of the design will go out of control at some point and will either cause delays in order to correct it or remain out of control to the detriment of the programme.
5. *Review and audit capability*. The ability to assess the progress (or lack of it) of the design. Throughout the progress of the programme the team and its management must have the capability to monitor progress in order to take the right decisions along the way. Along with the processes described above this means having sufficient discipline and authority to redirect or delay the programme.
6. *Management commitment*. The support from those with the power to make decisions throughout the programme. This is probably the most important enabler of all, because without commitment from management the programme is likely to suffer failure at the earliest signs of stress or difficulty. The commitment sought from management to achieve success demands the ability to recruit and train staff, to provide the necessary funding and commitment to capital expenditure and to provide the empowerment of the team and the individuals in the team to organize and manage the day-to-day programme activities. Without this the programme is doomed to an early failure.

The Structure of the Design Process

The design process is a complex picture; that is to say, what happens in people's minds and in the collective thinking of teams is difficult to comprehend. Yet if we are to improve the outcome of the design process, the design itself, we must first of all understand and be able to measure it.

Referring back to Sec. 1.2 where the composition of design is defined, there were five elements to consider:

1. Define the requirements of the customer.
2. Generate ideas by innovation.
3. Understand the function.
4. Translate the design into drawings or a design file.
5. Transfer the information into manufacturing.

These five elements translate into a structure that can be thought of as:

- The translation of customer's requirements into design criteria (elements 1 and 2 above)
- Understanding how the design functions (element 3 above)
- Converting the design into hardware, building and testing it (elements 4 and 5 above).

Although they occur sequentially and can be matched with the management process on a base of time, it should be understood that there are often feedback loops in the process described above, which means that some of the design activities can be repeated, albeit at a different level of detail in different phases of the programme.

Let us look at the content of each of these sections in turn.

Translating customer requirements into design criteria The objective here is to blend the known requirements of the customer with the best engineering approach to meet these needs. This demands that the level of risk (in engineering terms), the availability of new technology and the type of marketing introduction are known. For example, a product needed in the market-place quickly would not be a suitable candidate for a new technology because this would carry a significant element of risk in terms of time-to-market. By the same token, a product idea that demanded the use of a new technology would need a longer time for development.

Fundamental to the process at this point is that the customer requirements are clearly and unequivocally defined, and this is not always an easy thing to achieve. A company must decide firstly whether it is to be a leadership company or one that follows the competition. This is not to say that one of these is better than the other, just that it is important to decide which. Secondly, it should be understood that the customer does not always know best what products are required. A good example of this is the Walkman personal stereo system. If customers had been asked prior to the common introduction of these devices whether they wanted a music system which they could take with them anywhere (even jogging!), and use headphones to provide the sound delivery system, I am sure they would have rejected the idea on the basis that it was unnecessary and headphones were outdated. Few products recently have enjoyed such success as these gadgets. Similarly, Post-It notes were introduced without the customers' consent and I defy anyone today to find an office desk that does not have one of these stuck to it or a pad of them for future use. What would we do without them?

Suffice it to say that the customer cannot always be relied upon to tell you which products are required in the market. Distinguish clearly, however, between this and the fact that once the product is chosen in broad terms it is the customer who needs to tell us what it is in the product that gives satisfaction.

The Kano model in deciding what satisfies customers, the designer must be aware of three separate areas relating to customer requirements. These areas represent individual aspects of customer satisfaction which do change with expectations and time. They are:

1. *Basic customer requirements.* These are the kind of requirements that customers have, yet rarely express. They are the 'expected' requirements which are assumed to be available by the customer. For example, a customer buying a car in Europe is unlikely to ask for a heated rear screen as a specific need. The customer will expect there to be one fitted but will not request it. Contrast this with some years ago when such items were extras in our cars.
2. *Performance related customer requirements.* These are the needs that the customer is likely to express at the time of enquiring about a product. Continuing the analogy with a car, the customer will ask for a particular performance from the engine in terms of acceleration and speed and may request wide wheels and disc brakes for better handling and performance. These requests are all for performance related aspects of the product which the customer knows about and would like to have. As time goes by and as expectations increase some of today's needs may be regarded as basic in time to come.
3. *Exciting customer requirements.* These are customer requirements that the customer is unlikely to request simply because they will normally be outside their range of knowledge or vision. As a result any additions in this category are likely to cause excitement because they will be seen as new and an advancement to the product as known before. It is difficult to quote examples of this type of exciting requirement because they quite rapidly become performance related or basic. Currently, such features as heated seats, heated front screen and heated mirrors are regarded as exciting by the ordinary motorist.

Kano plotted these three types of customer needs on a graph of customer satisfaction versus degree of achievement (Fig. 1.6). We can see that in the case of the basic requirements, failure to achieve the need will cause

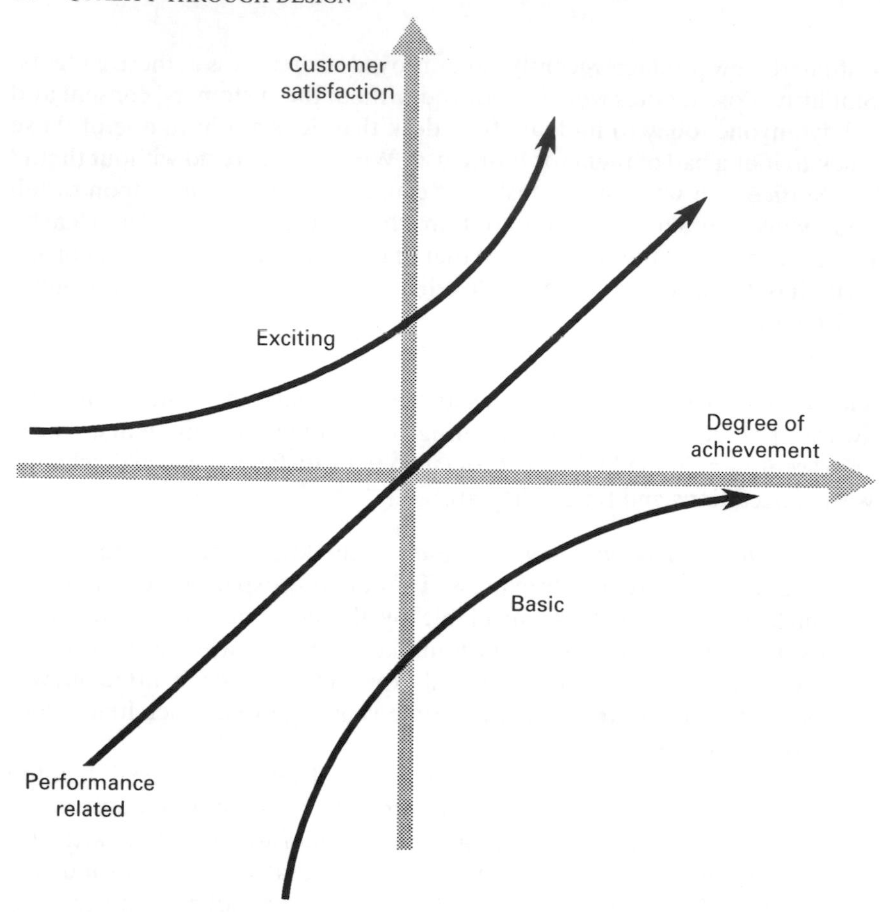

Figure 1.6 The Kano model.

customer dissatisfaction and only the full implementation will provide anything close to positive satisfaction for the customer. In the case of performance related requirements, the customer will be dissatisfied if we fail to provide them, but will become satisfied to a fair degree once a significant level of performance related requirements is achieved. The exciting requirements, on the other hand, will always add to customer satisfaction. The more of these features that are achieved, the higher will customer satisfaction rise.

The message is clear. Basic requirements are a must. Performance related requirements bear a direct relationship with customer satisfaction and any exciting requirements will serve to boost this satisfaction even higher.

Obviously market research is a key activity in defining what product is required. This will involve an analysis of the market, an analysis of the demand and a study of any new product attributes versus barriers to entry that might exist. The input from marketing to help to define customer requirements will need a clear definition of the following:

1. A description of the market itself to show how and what the product will be used for.
2. What the quantities expected are for the product.
3. What the expected selling price is and how it translates into product cost.
4. What the segmentation of the market is. What the variants of the product are that can be sold, perhaps with different profit margins.
5. What method of sales will be employed—direct sales, through dealerships, mail order.
6. What the competition is in this area of marketing.

This is marketing input, and it is marketing that is expected to initially define the need for a product and estimate the business worth of the product. Once this has been done the detailed customer requirements have to be established. This is a most important part of the process and cannot be addressed lightly because if customer requirements are not accurately and fully established at the outset of the design, major disruption in the design process can occur later, which may incur both schedule and cost impacts to the programme.

Customer requirements are not easy to come by. The 'voice of the customer' is a precious commodity and quite often can be at variance with what the marketing organization regard as customer requirements. The Japanese take great pains to identify the 'voice of the customer'. When Honda were embarking on the design of small trucks for the American consumer market they wanted to be sure to get the true customer's requirements. To do this they did not just go to the marketing and sales organizations, but actually took steps to get a more direct contact with real customers. They sent a team to investigate the situation in the car park of Disneyworld in Florida. This team waited for truck owners to return to their vehicles and then they interviewed them on the good and bad points of the truck that they owned. At that time the US automobile industry had a virtual monopoly on the production of these vehicles. However, the exercise of getting existing customers' opinions on real customer perception enabled them to produce their first Honda truck, which was a winner as soon as it was produced.

What the design engineer needs at the early stage of design is a full description of customer requirements for the product. In general terms this description should include the following quantified factors for the product:

- Performance parameters expected, including limits. The amount of detail here obviously depends on the type of product. One can visualize the difference in the size of this list for a motor car versus, say, an electric shaver. In essence the detail should cover the sort of information one sees in the brochure that advertises the product.

- Reliability
- Cost
- Weight
- Volume
- Production quantity
- Life cycle
- Safety and legal requirements
- Noise level
- Schedule
- Sales and service strategy

The next stage in the process is to translate these requirements into product goals. Here a negotiating process is initiated which iterates and reiterates the alignment between each customer requirement and its equivalent supplier specification. To distinguish between these two, imagine a customer requirement for a product to be as quiet as a refrigerator. Then the supplier specification would state that the continuous noise level must not exceed 35 dBA. This would be a fairly easy one to reconcile, but where there are high demands in the customer's requirements a good deal of negotiation will need to take place. In any event much has to be done to make the transition from customer requirements to product goals.

Product goals tell the design team in essential engineering terms what is required in the product. Before effective design can commence it is necessary to specify how the product or system is to work. The preparation of this information is a singularly engineering activity (just as understanding the market was a marketing activity). It requires somewhat more input data than just the product goals. In deciding how the product will work one needs to take decisions on such things as risk and competitive influence. Part of the study at this stage of the process is to assess whether there are any new technologies that are becoming available that can meet the functional requirements of the product and also whether there are any new ideas for embodying an old function that may apply. If there are candidates in either or both of these areas, an assessment of how much risk this might impart to the design must be made. Further, this must be weighed against the time-to-market requirement and the current assessment of competitive elements in the market.

The functional specification is the foundation upon which the design will be built. It should state in explicit detail exactly how the product will function. In its final form it is the document that will be used by every designer and engineer on the team and as such will be used to integrate activities as well as highlight areas of difficulty in achieving the functional performance. The document is best prepared by a systems engineer (or systems engineering team, depending on its complexity) who has a full and overall view of the product system and its operation. As the document develops, it becomes a rule-book for the engineering team and a statement of intent to the product planners and marketing personnel. Any discrepancy from customer requirements which has escaped the original negotiation will through this document be highlighted anew. Changes are expected to be minimal, and indeed the document is so important that once issued any changes should be the subject of discussions at a high level between engineering and marketing.

A design team cannot proceed successfully unless the functional specification is developed, completed and agreed between all parties as the final description of the requirements for the product.

Understanding how the design functions Once the requirements of the customer have been clearly understood and translated into terms that the engineer can relate to then the part of the design that most people recognize as 'designing' can begin. This means bringing together ideas that enable a product to satisfy the requirements of the customer and processing all those ideas to ensure that they meet all the elements of the design goals, such as reliability, safety, etc. This part of the design process culminates in the preparation of design drawings or files which are the designer's translation of the input material and ultimately go forward to enable the product to be manufactured.

This part of the process requires that a number of enablers is in place. Failure to complete the process of understanding function, because one or more of the enablers is either not in place or not properly implemented, will inevitably cause shortfalls in the quality of the ultimate design. Quite often, I believe, a design goes ahead simply on the basis of having a skilled designer prepare drawings which seem to adequately represent the requirements. The results of this approach are, at best, a design for which a significant reiteration is required to eliminate the errors and, at worst, a design that fails to meet fundamental requirements, culminating in the need to redesign the product. The lack of attention to addressing the understanding of function inevitably causes delays in the delivery of the product and often major problems for marketing and the quality image of the company. Figure 1.7 shows the main enablers for this part of the design process.

There is a primary transition required to embody all the requirements of the previous phase into design data. It is important that this is done carefully and methodically with proper documentation to back up the process. It is essentially the job of the systems engineer to ensure that this is carried out professionally and fully. There will be alternative approaches that must be evaluated on a logical basis. There will be problems of interpretation and integration which will need to be ironed out to give an optimum compromise. Many decisions will need to be made during this part of the process and these will need to be continually checked and justified with respect to the

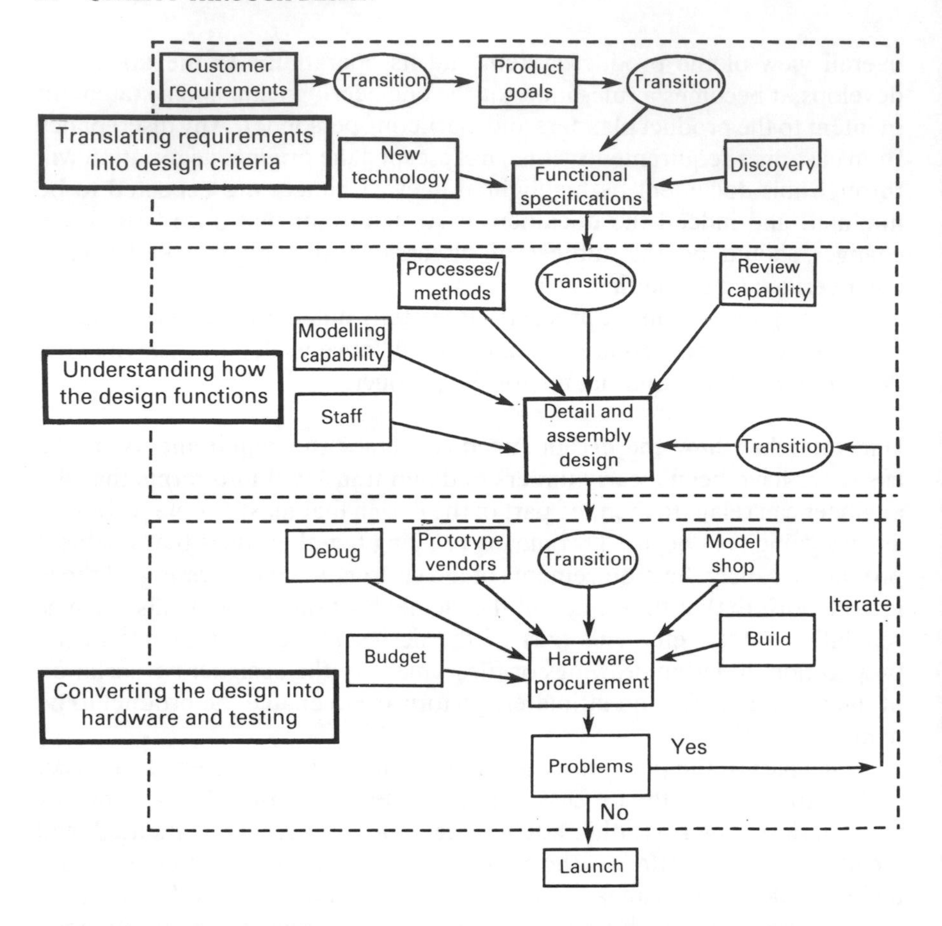

Figure 1.7 The design process.

total system. Staff must be available in adequate numbers to achieve the task and must have the right training, capability and skills mix. Experience is the best test of whether staffing is right or not.

There will be a need to have the capability to model, analyse or simulate parts of the design. This does not mean that a sophisticated mathematical treatment is necessary, but some level of modelling capability, however small, will be required and this should not be overlooked. Of course, with a product that has a high technology content this capability will by necessity become sizeable and will benefit understanding of the design, however small or large its need.

There will be a need to have processes and methods in place to support this phase of the activity. The type of process, such as critical parameter management, will create the central theme of the design itself. However, there will also be those processes that enable the design team to integrate their efforts, for example an agreed process for reference datums throughout the design, and there will be a need to specify tools and methods such as computer-aided design and the standard processing programs that go with it. All these will be required during this phase and must be in place for the start.

Throughout the period of understanding the function of the design there will be a need to review progress in all areas. This can be done at the working level by implementing peer reviews (see later in Sec. 4.4) and at the management level by more formal reviews to enable the right decisions to be made quickly and without disrupting the design process.

Finally, there will be a need to accommodate into the design the updated information that comes from testing and evaluation of the hardware. This will be in the form of a transition of the solutions to problems during the iteration of the design itself. It is this iteration cycle that we seek to minimize. The proper attention to functional understanding and full commitment to all the enablers will serve to minimize this 'design by iteration'. It will also enable experience to be captured and with proper documentation of the data will help to further reduce the design cycle of subsequent products.

Converting the design into hardware and testing The first thing to say about this part of the process is that there should be no rush to implement it. This is where large amounts of money start to be spent. It is not economically effective to find problems by testing the hardware; neither is it economical to solve problems by testing their potential solutions on hardware. The review process mentioned in the previous phase should certainly have the capability and be used to recommend when the right time arrives to build the hardware. If there is any doubt further work in the previous phase to understand the functional questions should be undertaken before embarking on building hardware.

Once again the main enabler for the procurement of the hardware is the transition from the design itself. The design team must be confident that they have a design that is robust, has adequate latitude of function, and is understood in terms of its manufacturability and sensitivity to the critical parameters of the design. Confidence in the design thus far may, under close scrutiny through the review process, suggest that hardware should be procured. This will require a number of enablers for this to be in place.

Is there an adequate budget to buy the hardware? If parts are to be made by outside suppliers, do these vendors have the capability to produce the parts using a 'soft tooled' process in line with the assumptions made in the design? Is there a need to provide support from the model shop to bolster the supply of parts and is this available? Are the people in place and have the processes been defined to enable both the assembly of the parts and the debug of the hardware once it has been assembled? All of these questions must be addressed to ensure that the procurement process results in quality hardware.

The primary purpose of the hardware at this stage is to simulate use of the product by the customer in order to highlight any problems that must be resolved before the product can be launched on to the market. The exact test programme will be dependent on the product itself, but it is important that testing explores all the extremes of the use of the product and therefore stress testing, that is testing at the extremities of certain conditions, should be employed as well as testing at the more nominal conditions. Life testing must also be part of the test programme.

There will be a need to organize the data so that it can be used both to identify the problems and support the activity of developing solutions. This will involve more processes for managing the data and the activities that evolve to improve the design. It is this activity that will enable information to be organized for the next design iteration.

The phrase 'transferring information to manufacturing' is probably not the ideal description of the process. Manufacturing should have been involved fully in the whole process of design so far. Therefore, when the design has reached the point where manufacturing becomes directly involved in the creation and assembly of parts, there should be no surprises or questions as to how this should be done. Of course, there needs to be a formal process to actually 'issue' the drawings and specifications into the manufacturing environment, but this should only involve the administration and control of the technical information to provide a formal configuration of the product. Often manufacturing will want to use the early or preproduction stages of the manufacturing process to develop assembly techniques or operator instructions and this should be carried out with the full cooperation and involvement of the designers and engineers.

This transition towards manufacturing will almost inevitably generate the need for changes in the design itself, and for this a formal change procedure must be in place to enable full control to be kept on the overall configuration of the product.

1.4 THE EVOLUTION OF TECHNOLOGIES

There are two ways in which change can take place. One is revolution, where drastic steps are taken to effect a change overnight. The other is by evolution, where the philosophy for change is by a gradual process of changing one small element at a time. This applies to all aspects of the world we live in. We see revolution most dramatically in political changes and evolution most successfully in nature. We have seen that revolution causes the change to be abrupt, often violent and sometimes incomplete or certainly less than ideal. Evolution is seen to be more effective in its result and nature often shows it to be a perfect outcome.

In design we can choose to take either of these approaches, depending on what our objectives are in terms of time for the change and desired effect of the change. In cars, for instance, we see generally a process of evolution where just a few elements are changed from the previous model to make the car look like something new. Of course one essential change to be made in this case is a change to the appearance, and this can either be done by a major change to the body styling or just some features of it. These kinds of change are of course closely allied to the marketing strategy and represent a means to simply sell more of the products.

In more fundamental terms we have to look at changes that occur in design due to the genuine desire to make improvements in performance or safety or cost. Look, for instance, at the evolution in the design of a tin-opener. This has gone from the device that pierced the tin and then opened it by levering the cutter around the rim, to a rotary cutting disc driven around the tin's circumference by hand, to a modern electric device that clinically detaches the lid and then retains it magnetically. This is an example of how new ideas can improve designs and represents in the first instance a revolutionary change (levering to rotary) and secondly evolution of an existing manual design to a version that is electrically driven.

There are examples of both revolutionary and evolutionary design changes that have not been successful. Everyone has something that they hate because it is not as good as the old one used to be. This is often the result of poor use of a new material to replace a more costly one or a failure to actually understand the functional changes being introduced by a new idea.

Quite often, changes in design are brought about by need, need to extend the capability of a device, need to compete with another product, need to reduce costs. If we look at man's attempts to explore the world, we see in the early days the use of a boat driven by manpower—a small boat paddled or rowed by the traveller. As bigger boats were needed and greater distances were travelled, this form of motive power had to be increased. This was done in the obvious way by adding more manpower, resulting in galley ships with innumerable oars working in unison and pulled by slaves—evolution. The discovery of sail power transformed this scene dramatically and soon bigger, faster boats were able to travel long distances, wherever the winds blew—revolution. Soon of course rivalry in world trade demanded that these boats should become faster to enable the traders to speed up their trading throughout the world, and so more and more sails were added to get more and more power from the essential trade winds. The limits to this form of propulsion, once reached, encouraged the search for alternative forms of power to be brought into service. With the invention of the steamship the death knell of sail was sounded.

The study of the history of boat propulsion is fascinating in terms of development of design and technology (Fig. 1.8). It shows how changes are driven by need, limited by the generation of new ideas, and have no foreseeable end.

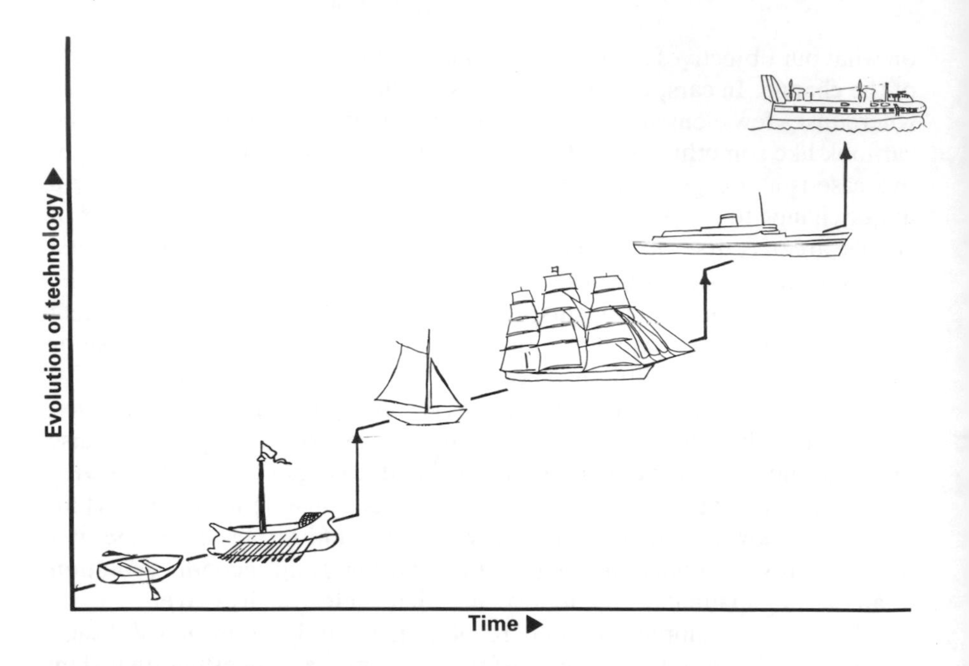

Figure 1.8 The evolution of technology.

The way in which we, as designers, approach change is important. We are all driven by our own ideas and biases. As designers we value the innovative approach to solving a problem and almost always feel that our own idea is best. However, there are dangers here. New inventions are rare. They are almost always made possible by the advent of some new material or other breakthrough which lies in parallel with the invention. For example, most innovations in motor cars have come as a result of new materials or computer science. The jet engine was made possible by the development of high temperature alloys and the computer itself by the development of thin film technology.

Therefore, given a technology, any invention to replace it, although it will look good at an early stage, will more than likely be equivalent to what already exists unless it has some enabling novel features to support it. So often I have seen one idea, which has been under development for some time, discarded and replaced by an alternative idea which had not been developed at all. The reason was always that the underdeveloped idea always looks better in the early stages and only when it has been fully explored are all the associated problems revealed. Reinvention of the wheel, a metaphor so often heard in design circles, reflects the practice of discarding what has gone before to use what appears to be a brilliant new idea. Even teams can fall into this trap just as easily as individuals. Instead of analysing existing data and applying their design expertise to evolve and improve it, they go ahead with a 'clean sheet' design as an alternative.

The process of critical parameter management described later in this book (Sec. 3.5) should help this practice to be largely eliminated. By understanding the functional operation of a technology and the critical parameters that drive that function, the designer should be able to see both how to extend that technology or why and how that technology may have reached its limit of extension.

1.5 CULTURE

Recently the Western world has looked towards Japan for leadership in terms of quality and productivity. We have seen Japan grow towards a nation of high quality, high technology goods which has swept the world and established new standards of innovation, design and manufacturing methodology. Many of the ideas in place in modern industries today originated in Japan. Apart from emulating their methods how can we learn from their culture to become more competitive?

Firstly, are we very much worse in the way we bring products to market than the Japanese? The curve shown in Fig. 1.9 has been published many times in comparing Japanese product delivery with that of the West. What it purports to show is the number of engineering changes in the product delivery cycle of a typical Japanese product compared with that of a US company. It shows the number and rate of engineering changes in a product from concept to launch. The significance in measuring the number of engineering changes is that every change represents a cost of non-conformance in the design and the accumulation of these and their phasing in time represents the quality of the design and subsequently the cost of quality. The Japanese curve shows an early rise in the change rate which peaks at the time of completion of the testing of the first prototype. Changes after that begin to reduce until, as the product comes close to launch, the changes to the design are at a very low level. By comparison, US companies are shown to be much slower in the early stages of development in identifying the changes required. The changes increase gradually through the testing of two or three prototypes and as launch approaches, because changes are restricted due to a need to support production, there is a slight decrease. Once the product is launched and operating in the field, the customer identifies problems and this promotes changes which have to be implemented in the post-launch period, and this causes the rate to rise again. The overall level of change does not subside until more improvements have been made to the product as a result of field experience.

This hypothesis raises some interesting questions about the differences between Japanese and Western ways of doing business. Firstly, how do the

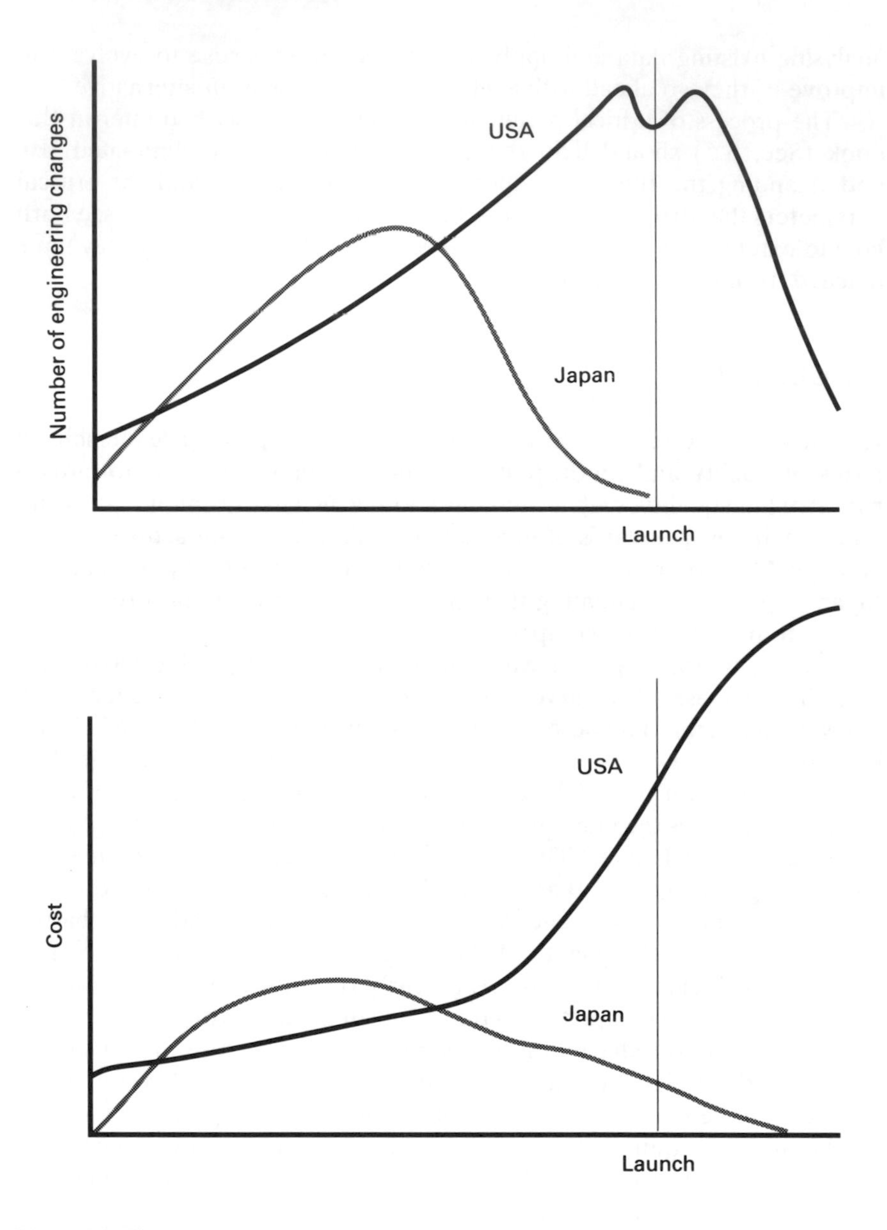

Figure 1.9 The cost of engineering changes to launch a product.

Japanese find it possible to develop a product with only one prototype machine and yet are able to identify shortfalls so early? Why do the US and Western companies have to respond to customer feedback in the post-launch period?

The popular consensus seems to be that the Japanese work harder, have more stable investment in their industries, are better equipped in manufacturing and somehow seem to have found methodologies that work in the areas of design and quality control. However, there are many cultural differences which also contribute to their success.

In the first place they have a completely different market-place and set of customers. Admittedly they do pay careful attention to customer requirements and no doubt this is a strong part of their success, but once they are committed to a product they tend not to change it immediately to satisfy some new requirement identified by the customer. What they do take care to do, however, is to log the requirement for a change and make sure that they introduce this new requirement in the next product. Their Japanese customers are also more tolerant to this and other inconveniences. They have more respect for the machines they use and if they break down they are more inclined to blame themselves initially than is a typical Western customer. They seem to be able to get away with a greater involvement between the customer and the product than their Western counterparts. For example, on a copier product designed in Japan there was a functional part of the paper path that protruded below the bottom of the frame of the machine and would be likely to be damaged in the course of transporting the machine. When asked about this and about how a fork lift would be likely to damage the mechanism, they replied that the fork lift drivers would be instructed where to place the forks so that damage to the crucial mechanism would be avoided. Thus they have great expectations from their people in looking after a product, whereas the West usually has to implement design solutions to protect fully against a possibility such as this. Quite often they test their products in the Japanese market before they introduce a similar product to the Western world.

If we look at the Japanese education system for their designers, engineers and managers, we find that it is not too different from ours. In general the technical people enter the company just as ours do after graduating from a university or college. They are, however, trained as design engineers and this means that they do the complete job of designing. They design, complete the engineering analysis, do the drawings, build the hardware, run the tests and analyse the data. Their knowledge of engineering tools such as QFD,Taguchi methodology, FMEA, etc., is learned during their education at a university and not during the training with the company that they work for. Additionally, when they do join a company they are given on-the-job training by an experienced engineer who is able to pass on experience and wisdom to the new recruit. This is regarded as a very important aspect of post-university training. Other training, more in line with Western ideas, is given also in the form of off-the-job training using seminars and lectures, etc.

Tracing the path of the decision process reveals some significant differences between Western and Japanese management. Looking first at a

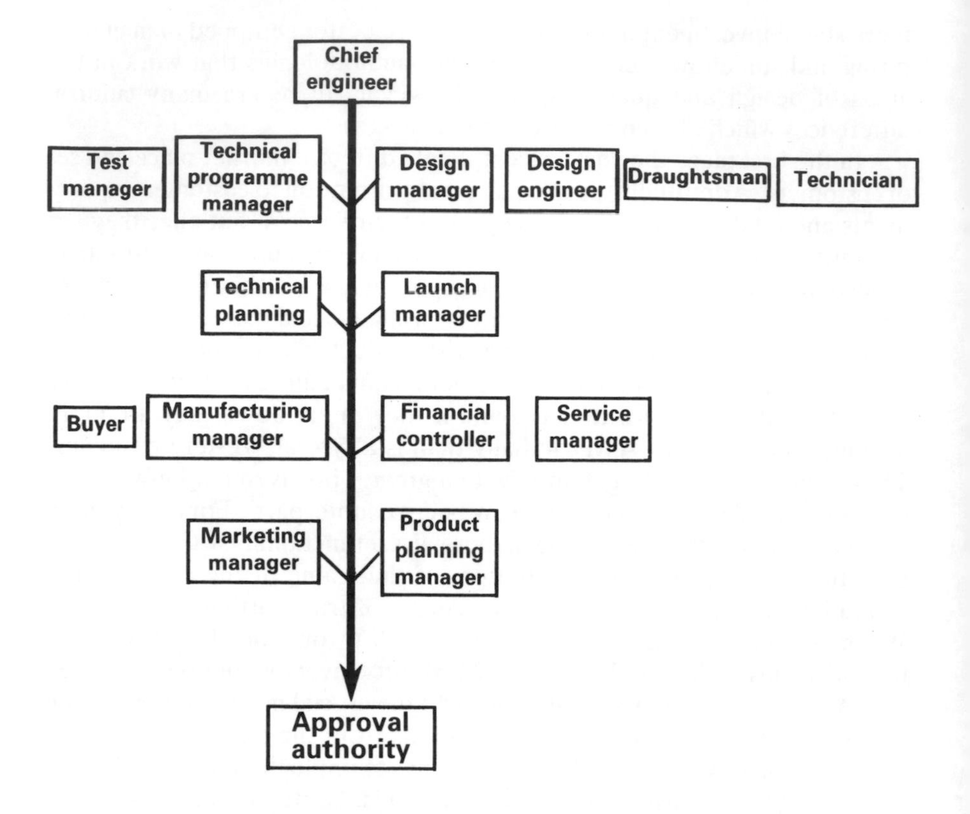

Figure 1.10 Western organization.

typical Western working style (Fig. 1.10) we see that the chief engineer gathers a technical programme manager, design managers and all the other associated technical staff with which to come to a consensus on decisions for the programme. Approval of these decisions will quite often pass through the marketing, manufacturing and technical planning functions before it reaches a top level of management for final approval to be made. In the case of a Japanese programme team (Fig. 1.11) a group programme manager will reach consensus with all functions—marketing, manufacturing, quality, servicing, financial—and then have direct access to top management who can then give approval on consensus opinion. The fundamental difference here is that top management decisions in Japan are brought about by dealing with options that have been generated through consensus, rather than bringing the various arguments from each of the functional areas for top management to arbitrate on.

This drive towards consensus by the Japanese is effected both by co-location of the personnel and by having close access to top management at all times. Design teams are invariably housed in one large room with no

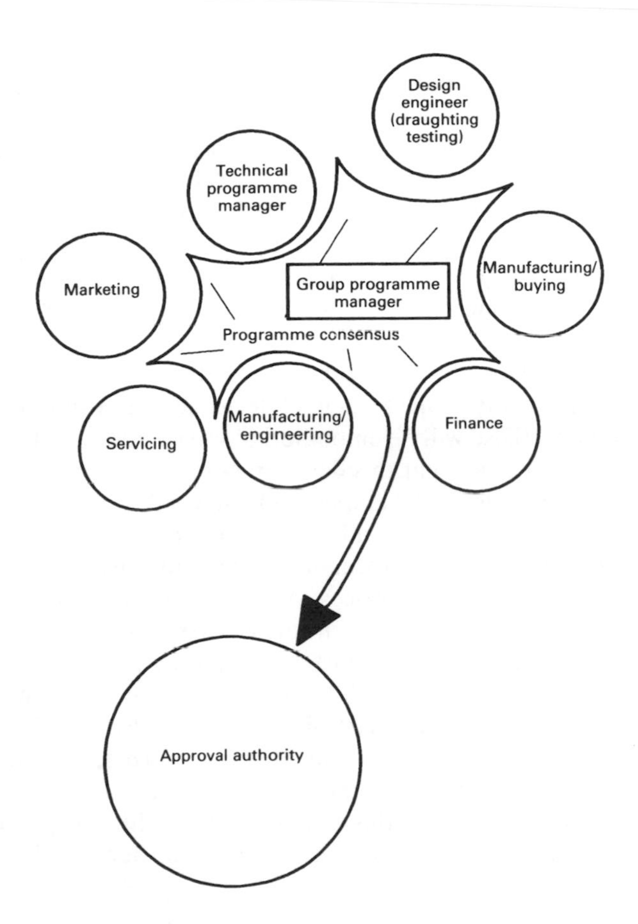

Figure 1.11 Japanese organization.

partitions dividing the personnel. In this way leadership is more readily established and a greater communication is achieved, reducing the difficulty of ensuring that the systems problems are appreciated by all parties. Top management, the level that makes the decisions on the investment of the company's money in new projects, is only one step away from this team organization. In other words, the organization is more horizontal and flat than the equivalent Western one.

The Japanese also tend to manage more by fact. This means that they allow the facts to be presented to their superiors rather than their conclusions from the facts. This does not mean, however, that decisions are not taken at the lowest possible level, but more that there is a greater sharing of information across the board so that system interactions can be more readily identified. The management-by-fact syndrome is manifested by the style of presentation seen by a Japanese engineer to the boss. It contains mostly diagrams, charts and graphs—the facts of progress. More likely a typical Western presentation will contain words and statements about results, conclusions and plans.

The Japanese engineer appears to be more aware of the need for a date as part of any technical objectives for the task in hand. It is as if an activity is never thought about without thinking of it in the context of a completion date. Moreover, having established a date for completing the task, a Japanese engineer is committed to that date and if there is ever any danger that it will not be mct, all possible efforts will be made to recover the situation.

There is another subtle difference in the decision process for a product. There is a point in any project when all decisions as to what the product will contain and how these will be embodied has to be addressed. It is generally thought that the earlier that this decision point can be reached, the earlier the product is likely to be developed and launched. This of course is true as long as the decision which is made is a sound and complete one.

The Japanese follow a process of negotiation, which they call Nemawashi (Fig. 1.12), which is a process of mulling around information among all the functions so that a sound consensus decision can be made. The word 'nemawashi' literally translated means 'preparing the groundwork', but in this context the practice is more like 'oiling the wheels' so that relationships are confirmed and opinions and biases fully explored. This often takes longer than the apparent equivalent Western process, but once it is made there is a much shorter run for the final goal, launching the product. One other important element to this is that Nemawashi, the process of negotiation and consensus, takes place against a background of clear and consis-

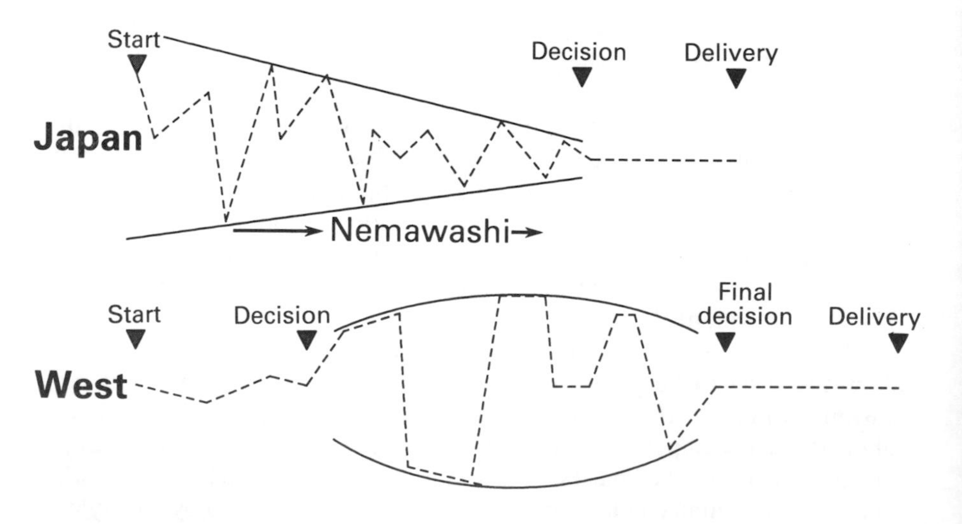

Figure 1.12 Nemawashi.

tent management strategy. This being so, it is almost inevitable that once the decision on the product is made it will meet the defined strategy for the company and will therefore get rapid and sound management approval.

By comparison, Western companies often take an early decision on a particular product strategy (that is they reach consensus at an earlier stage) but then go into a period of trial and error and adjustment to that decision which has the effect of slowing down progress significantly. One Japanese manager told me that if you liken product development to a game of baseball, US companies want to hit the first ball and make it a home-run, whereas Japanese companies select the best ball through total quality management and then hit it cleanly. It may not be a home-run, but by working from base to base, the game can be won!

So what does this all mean to product development in the West? Firstly, we cannot become exactly like the Japanese. We do not have their culture and so the changes we would have to make would be far too swingeing to be feasible. We are good innovators and have a good history and background of engineering which have brought many innovative and dramatic products to fruition. What we lack in general is the disciplines and attention to detail in the management process that the Japanese appear to be so good at. We need to take the best Japanese practices and tailor them to suit our own culture and style. We need to teach these and basic design engineering in our universities. This means a greater cooperation than there is at present between industry and the academic institutions.

Although many of our industries have been taken over and dominated by Japan in recent years, we have the wherewithal to reverse this trend and establish our own leadership and competitiveness in the future.

CHAPTER

TWO

THE DESIGN PROCESS

2.1 THE ARCHITECTURAL PROCESS

As customer's requirements become defined there is a need to study the alternatives of satisfying these requirements. One of the underlying foundations of a good system design is the optimization of the architecture supporting that system. What is meant by architecture in this context? Obviously highly complex systems have many elements that link together to build finally into the complete product. It is the arrangement of these elements both in geometric or spatial form and in operational relationships that comprises the architecture. The alternatives for combining these in terms of three-dimensional space are usually many, but the alternatives for changing the functional relationships are also often varied and many. Even in systems that are small there are usually enough variations to warrant the study of alternatives and the optimization of these.

The development process that ensues from this exercise is in many ways the outset of the concurrent engineering process that proves to be needed. It is at this early stage of architectural development that there is a need to bring together all the parties that will have an influence on the subsequent design. The process that follows is not the only way in which this can be carried out, but it is one that has been tried with success and that complements the succeeding design process.

The architectural development process flows logically through five broad phases:

Definition
Concept generation
Concept reduction
Concept evaluation
Output

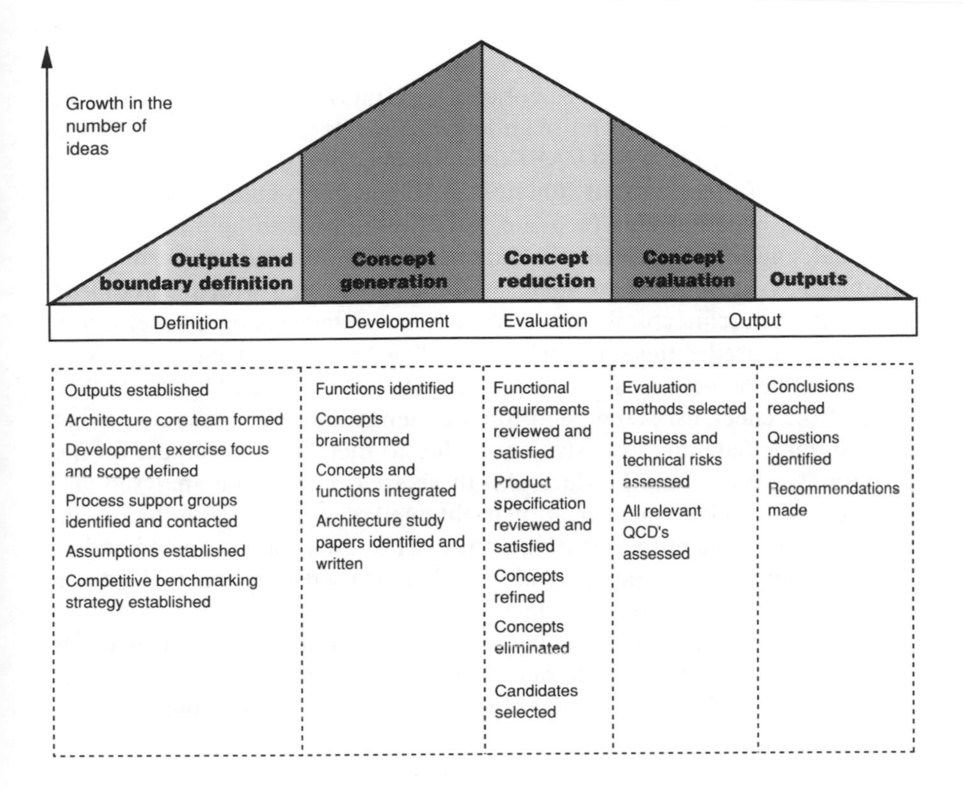

Figure 2.1 Architecture development process.

The content and general execution of each of these phases is described. Figure 2.1 illustrates graphically the activities involved in each of the phases and what their outputs are.

Definition Phase

The primary objective of this phase is to establish the outputs of the product and define them clearly. These of course are required to meet the customer's requirements. In addition to the nominal outputs, the boundaries of the outputs are also developed to establish scope and the assumptions that are made. At the start of this phase the architectural core team is formed, the supporting personnel are contacted and clear lines of communications are established with the process support groups. The strategy that the programme will use to develop the optimal architecture is also defined.

Concept Generation Phase

As the team progresses into the concept generation phase, the quantity of work gradually increases as more and more possible alternatives are formulated and defined. The use of FAST diagrams (see Sec. 3.3) is recommended here to identify the relevant functions of the system. Once this has been established, brainstorming (see Sec. 3.9) can be used to help to generate potential solutions that meet the functional requirements and satisfy the customer. As new ideas are generated and integrated with the required functions, the architectural concepts will begin to take shape. A useful tool that has been used at this stage in the process is a check-list which provides a prompt of the subject areas that require review (Fig. 2.2a). One of the key parts of the check-list is the need for the identification of those areas that require immediate and intensive study due to their level of impact on the functional output of the product. All these areas have to be analysed and studied so that a clear quantification is obtained regarding their influence on the operation of the proposed architecture. The new information and understanding which these studies provide can then be used to update the existing concept proposals or form new ones.

Growth in the number of ideas and concepts is substantial through this phase and attempts to filter them at this stage should be suppressed as the objective of the phase is to accumulate as many ideas as possible.

---

1. *Product theme.* Ensure that an understanding of the product theme exists. This should guide the top level concept development decisions.
2. *Product specifications.* Ensure that a clear product development specification exists. Review all concepts against this specification.
3. *Critical design issues.* Identify any important design issues, particularly those with systems level effects. Analyse these design issues to whatever degree is necessary to enable the required decisions to be made. Document these findings in an architecture study paper.
4. *Competitive benchmarking.* Be aware of what the competition is doing and how they are doing it.
5. *Relevant standards.* Be aware of the important standards that must be satisfied by the concept under development.
6. *Establish the minimum level of detail* required for each concept to ensure that a fair evaluation can take place.
7. *Preliminary evaluation criteria.* Begin to record evaluation criteria as they become clear. Also begin to think about evaluation processes, weighting techniques, etc.

Figure 2.2a Concept generation check-list.

Concept Reduction Phase

So far we have seen the number of concepts rising as more and more ideas are generated. The objective of the concept reduction phase is to start to reduce the number of ideas by eliminating those that are considered to be inferior. Thus a large number of separate functional and conceptual ideas is condensed into a workable set of candidates by careful filtration and selection. The process of reduction of ideas can be applied in a number of forms. Sometimes, a simple decision to reject an idea can be made simply because a key parameter is not satisfied and no feasible way can be found to resolve this. Where there is a need to compare similar alternatives by way of elimination the Pugh process (see Sec. 2.2) of selection can be employed with favourable results. Sometimes it may be necessary to reiterate back into the concept generation phase as the very processes used to reduce the number of selections quite often generate new ideas as a result. Other reasons for this are:

- The team's understanding of the existing concepts is enhanced.
- Areas that require more detail to enable proper evaluation are identified.
- New ideas appear as existing ones are reviewed.

At the end of the concept reduction phase, the remaining candidates should be as complete as possible and must be developed to the same level of understanding as one another.

Here a word of warning is in order, and this applies at any stage during the design process. *All designs, concepts or ideas that are immature in their development and understanding will invariably look better than the equivalent mature ones!*

Concept Evaluation Phase

During this phase an explicit, well-defined method of evaluation and decision analysis is required for the comprehensive evaluation of each of the candidate architectures. Two methods are recommended. One is the Pugh method which relies on a large number of unweighted criteria and a simple unscaled evaluation of the concepts relative to a preselected reference datum. The second is the Combinex method which is similar but features weighting and acceptability or preference curves for the criteria. This allows relative differences in importance between criteria and the relative difference between concepts against each criterion to be captured in the evaluation scoring process. In general, the Pugh method is better when there is limited quantified data available, whereas the Combinex method can handle numerically quantified criteria if these are available.

Both methods require suitable evaluation criteria. Throughout the early phases of the development exercise the concept generation check-list offers reminders to record evaluation criteria at any time they are thought of. In

1. *Relevant product specifications.* Ensure that the relevant product specifications are represented in the list of criteria. The relevant product specifications should come from the customer's specific requirements either directly or indirectly. Qualitative rather than quantitative evaluation emphasizes relative comparisons between concepts and therefore real numbers to represent the specifications are not absolutely essential.
2. *Relevant architectural standards requirements.* Ensure that any relevant architectural standards requirements are represented on the list of criteria.
3. *Technical risk.* Consider assessing technical risk as part of the evaluation process. How dependent is the concept on a technological breakthrough? Assess technologies being used against technology readiness criteria.
4. *Business risk.* Consider using business risk as a criterion for selection. Establish patent positions, market acceptance, safety, etc.
5. *Concept advantages and disadvantages.* Utilize the unique advantages and disadvantages from each of the concepts to establish some of the evaluation criteria. This ensures the capture of both the good and bad points of the concepts in the scoring.
6. *Effect on the other parts of the design.* Attempt to identify any interactions that may occur between subsystems and include these as evaluation criteria. Some examples are: power, space, cost, accessibility.
7. *All factors having an effect on cost, quality or schedule.* Include as evaluation criteria any factors that will affect these programme requirements.

Figure 2.2b Evaluation criteria check-list.

this phase they must be put to use, but only after careful selection. Similar to the concept reduction phase, the large list of evaluation criteria must be reviewed and refined to yield only those that are useful in differentiating between the architectural concepts. Like the concept development check-list, a check-list to help with the development of evaluation criteria has been developed and this is called the evaluation criteria check-list (Fig. 2.2b). It lists several categories of developed criteria which may be relevant to the exercise and should be reviewed. Once the team has reached a sufficient level of satisfaction with the criteria selection, the list should be 'frozen' and definitions carefully written to eliminate ambiguity or vagueness.

If an evaluation method has been chosen that uses weighted criteria, a technique for arriving at the weighting must be selected. With the Combinex method, typically, a paired comparison technique is used, where each criterion is compared with each of the other criteria in turn and decisions are made as to which of the pair is more important in each case. Often, it is helpful if a carefully worded question is developed that can be asked whenever criteria are compared. For example, such a question might be 'Which of the two criteria has more bearing on customer satisfaction?' or 'Which of the two criteria presents the greater challenge to the technology development process?' Careful and precise wording is important as it can have a major significance on the answers compared with alternative questions like 'Rate the criteria against customer satisfaction' or 'Which criterion will be easiest for technology?' The team can and should conduct the weighting development effort together in order to reach a consensus for each pairing. Alternatively, team members can make their own evaluation and average the results at the end. The second approach will take less time; however, the danger of misinformed or biased decisions will have the effect of diminishing significant differences between the criteria. Once the weights are determined, they are normalized to percentages.

Acceptability or preference curves (happiness curves as they are sometimes known!) are a valuable feature of this selection method. These curves acknowledge the fact that not all relationships are linear. In fact, some have step changes or thresholds that we must be aware of. The preference curve allows a variable weighting to be applied and tailored to specific criteria. During concept evaluation, concepts are assessed against the specifically tailored curves developed for each of the criteria, the position on the curve determining the relative importance to be used in the final total.

It is recommended that a larger group participates in the evaluation sessions to ensure that all the important areas are represented in the criteria and the evaluation of the concepts against those criteria. This is particularly true if the criteria list is short. The shorter the list, the more weight will be lost by one 'missing' criterion.

After the main evaluation session, the concepts may be rank ordered according to the overall score or presented in some other suitably graphical way such as Pareto (see Sec. 3.9). A discussion of the results and a sensitivity analysis, if desired, should be included.

Output Phase

Conclusions and recommendations are developed by the team during this phase. The activity should include a check against the original objectives and any deviations accounted for. Whenever a critical design issue is addressed it is essential to document this in what could be termed an architecture study paper. This paper will be a full account and record of all assumptions, thinking and conclusions that went into the critical design issue discussions. The value of this documentation will be appreciated when, at a later stage, the issue is reopened, perhaps with some modified constraints.

Process Support Group

It is important that the correct supporting expertise is available to the study team throughout the architectural selection process. The needs here will of course vary widely from project to project. However, by way of a check, here is a list of possible support expertise.

Technologists
Product planning

Marketing
Competitive evaluation
Architectural standards
Mechanical, electrical, electronic design
Industrial design
Ergonomics
Manufacturing
Manufacturing technology
Service

Regularly scheduled interaction between all members of the team should obtain, as this is critical to the progress and credibility of the development activity. While the core architecture team should meet frequently, a weekly meeting with the support members is key to the timely information exchange in both directions. This helps to keep team members abreast of new developments and prevents unnecessary excursions outside the boundaries of the defined task.

2.2 IDEAS SELECTION

Throughout the design process there are always decisions to be made. Much of the time these decisions are in the form of choices. Which of the alternatives available to us will be the best for our requirements? Most of the time we make decisions on these choices as a matter of routine and most of the time this is a perfectly adequate process. Sometimes, however, it is necessary to adopt a much more formal process of decision making for a number of reasons:

1. We want to marshal our thoughts in a logical way to ensure that we have the necessary data to enable a good decision to be made.
2. We want to follow up on a decision and test it by modifying some of the data to see what effect it will have.
3. We want a means of documenting the decision process.

There are a number of processes that will enable the designer to organize data to enable a formalized decision to be made, but I have chosen two which I have found suit engineering design best, each being applied at a different point in the process for maximum effect.

The first method is by Professor Stuart Pugh and is a method that I find most effective when the subject matter is immature and not highly quantified. The method is therefore applied most often during the early stages of design, in the preconcept or concept phases, usually to choose a particular approach to a design solution.

The other method is known as Combinex. This relies on a fairly high

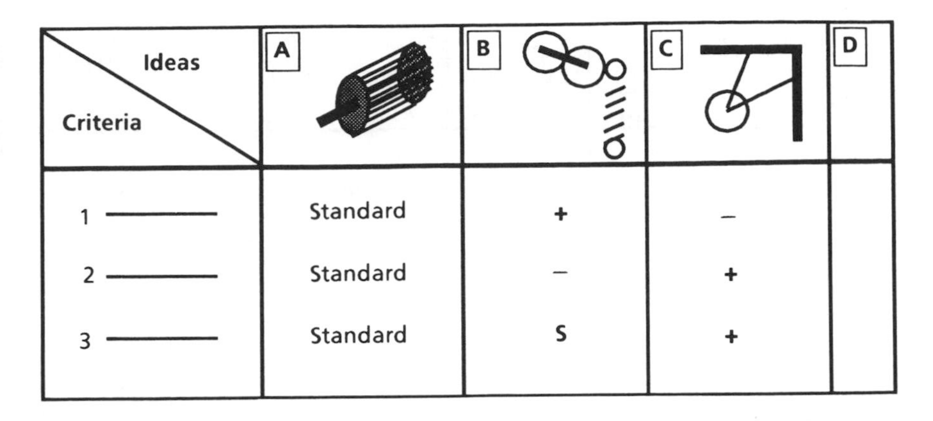

Criteria / Ideas	A	B	C	D
1	Standard	+	−	
2	Standard	−	+	
3	Standard	S	+	

Figure 2.3 The Pugh selection process.

level of quantified data and also a good understanding of customer requirements and their relative importance.

The Pugh Method

The team prepares the data using the type of chart shown in Fig. 2.3. An icon is used to describe each of the ideas being considered. This enables the team to visualize the idea and also form an image of what is in the idea rather than just relying on a number or a letter. It also enables the mind to flit from one idea to the next without difficulty. Judging criteria are listed on the left-hand column of the table. Then one of the ideas is selected as a standard against which each of the others will be compared. This is ideally something that is well established as a solution for the task and will have a good deal of knowledge associated with it. For each of the criteria a score of plus (+), minus (−), or same (S) is attributed according to whether the idea is better, worse, or the same when considered against that criterion.

At the end of scoring in this way, each of the ideas has the number of +, − and S totalled up so that comparisons can be made. It is at this point that the important work begins, because it is the comparison of the scores—both totals and individual attribute scores—that allows the mind of the assessor to understand the subtleties of the decision process itself. This brings a greater understanding to the comparisons between each of the ideas and usually causes changes to the original scoring. Going through such a process is sufficient to sharpen awareness of the attributes which allows the decision maker to make a much more informed decision.

One of the strengths of the process is that arguments break out within the team as the discussion goes on. These arguments are primarily due to a lack of understanding either of the criteria or of the concept itself. The arguments will often be the trigger for new ideas. By running through the

Ideas / Criteria	A Pulse width modulation	B Software to gradually release solenoid	C Fluid damper	D Magnetic damping	E Electromagnetic damping
Noise reduction	+	+	Selection	s	s
Cost	+	+	Standard	+	—
Time to implement	—	s	Standard	—	—
Performance	+	+	Standard	—	s
S	0	1	4	1	2
+	3	3	0	1	0
—	1	0	0	2	2
Selection		*			

Figure 2.4 Example of Pugh process.

process of discussing the matrix the team will discover just which elements of the ideas are and are not important to meet the selection criteria, and this in itself helps to focus the minds of the team members on those aspects that will eventually produce the best solution to the problem.

Example of the Pugh process The example shown in Fig. 2.4 relates to the selection of one of many proposed solutions for reducing noise in a mechanism driven by a solenoid. The mechanism driven by the solenoid suffered a very loud impact noise when the solenoid was released and it was postulated that if some control on the acceleration profile of the solenoid on release could be effected, this would significantly reduce the impact noise.

A number of solutions were proposed:

A Control of the solenoid release voltage by pulse width modulation
B Control of the solenoid release voltage decay profile by the use of software
C Use of a fluid damper to control release acceleration
D Control of the release acceleration by using magnetic damping
E Similar control using electromagnetic damping

The various solutions were defined in as great a detail as possible based on the knowledge surrounding them. In most cases this did not include cost or reliability information.

The selection process started by depicting each of the potential solutions along the top of a chart, as shown in Fig. 2.4. The criteria for selection of the best solution were written down the left side of the chart and were chosen as:

Ability to reduce noise
Estimated cost
Estimated time to implement the solution
Reliability of the device

The potential solutions were considered against one another and solution C was chosen as a standard against which each of the others would be compared. This particular solution was chosen because it was well known in its previous performance and application. Thus it would form an ideal basis for comparison of all the other contenders against the selected criteria.

For each of the potential solutions, the individual criteria were assessed against the standard as to whether they were better (+), worse (−) or about the same (S) and the symbols entered in the chart. Cost and reliability, for which there was no numerical data available, were assessed qualitatively against the standard, just as were both the other criteria.

At the bottom of the chart the comparisons were summed and the results studied. The first thing to look for is where there are little or no negatives (candidates B and C). Next those that had high positive scores were considered candidates A and B. This led to the selection of candidate B, the use of software control of the acceleration profile.

Quite often, of course, the solutions are not as obvious as this. When this happens the process is still useful in enabling an in-depth comparison to be carried out and in drawing attention to those factors that cause certain solutions to score low. Once this is understood the review team can search for ways of correcting the low scores and this will often lead to another alternative which may be better than the others due to the combination of some of the best points.

The Combinex Method

The Combinex method is much more complex than the Pugh method. It relies on a set of numerical data to describe each of the options under consideration and should not be used if these data are not available for all the options. An additional important point to beware of is that each option must be defined to approximately the same degree of understanding; otherwise it is usually found that the least defined option quite often looks much better than the others.

The method can be divided into seven steps:

1. *Definition of the selection criteria.* The first step is a very important step which actually reflects the customer's requirements. The decision maker must select all the criteria by which the choice will be judged. It amounts to listing every factor that will have an effect on the choice made however

important or unimportant it seems. Many of the criteria are fairly obvious and crop up regularly in most decision processes for design. These are cost, reliability, size, etc. Others are more obscure and relate to the specific customer's preference for certain attributes.

2. *Weighting of the selection criteria.* At this step, it is recognized that each of the criteria defined to make the choice does not have equal importance in the customer's mind. Decisions have therefore to be made about which of the criteria are most important to the customer and how they rate relative to the others. The best way to evaluate this is to assign cqual value initially to each of the criteria in percentage terms. Then look at each of the criteria and modify the percentage values, placing importance where the customer requires it and ensuring that the individual percentages add up to 100 per cent. Then do tests of reasonableness by comparing the weightings of pairs of criteria and modifying again until the whole relationship of weighting looks correct.
3. *Development of utility curves.* At this step we take each of the criteria in turn and ask what value of this criterion would make the customer 100 per cent happy. Then ask what value of the criterion would make the customer 0 per cent happy. Deciding on happiness percentages between 0 and 100 enables a curve to be drawn which is called the utility (or happiness) curve. A curve such as this for each of the criteria enables the decision maker to assign points which reflect the customer's satisfaction for any value of the criterion for each of the options.

   These curves (Fig. 2.5) help to describe the thresholds of satisfaction for the various criteria. For instance, with cost there is usually an upper limit above which a customer would not buy the product. Below that the customer's satisfaction will increase as cost reduces, but this may not be a linear relationship and below a certain point the customer may not care too much if the cost is lower. Other criteria may form a stepped curve as each change in the parameter will have a distinct level of desirability. Some factors may have sharp cut-offs which indicate that a value above a certain figure is totally unacceptable.

   The generation of these curves provides a tool to actually quantify customer satisfaction.
4. *Assignment of points to each alternative.* Now comes the activity that begins to measure each of the alternative choices. Each alternative is assessed in turn against each of the criteria and using the utility curves is assigned points.
5. *Application of weighting to the points.* The weighting of each criterion is now applied by multiplying the points given for each criterion by the appropriate weighting factor. This gives the true value of each of the options against the customer's requirements. Totalling these weighted points will now give the resulting assessment values for each of the options.

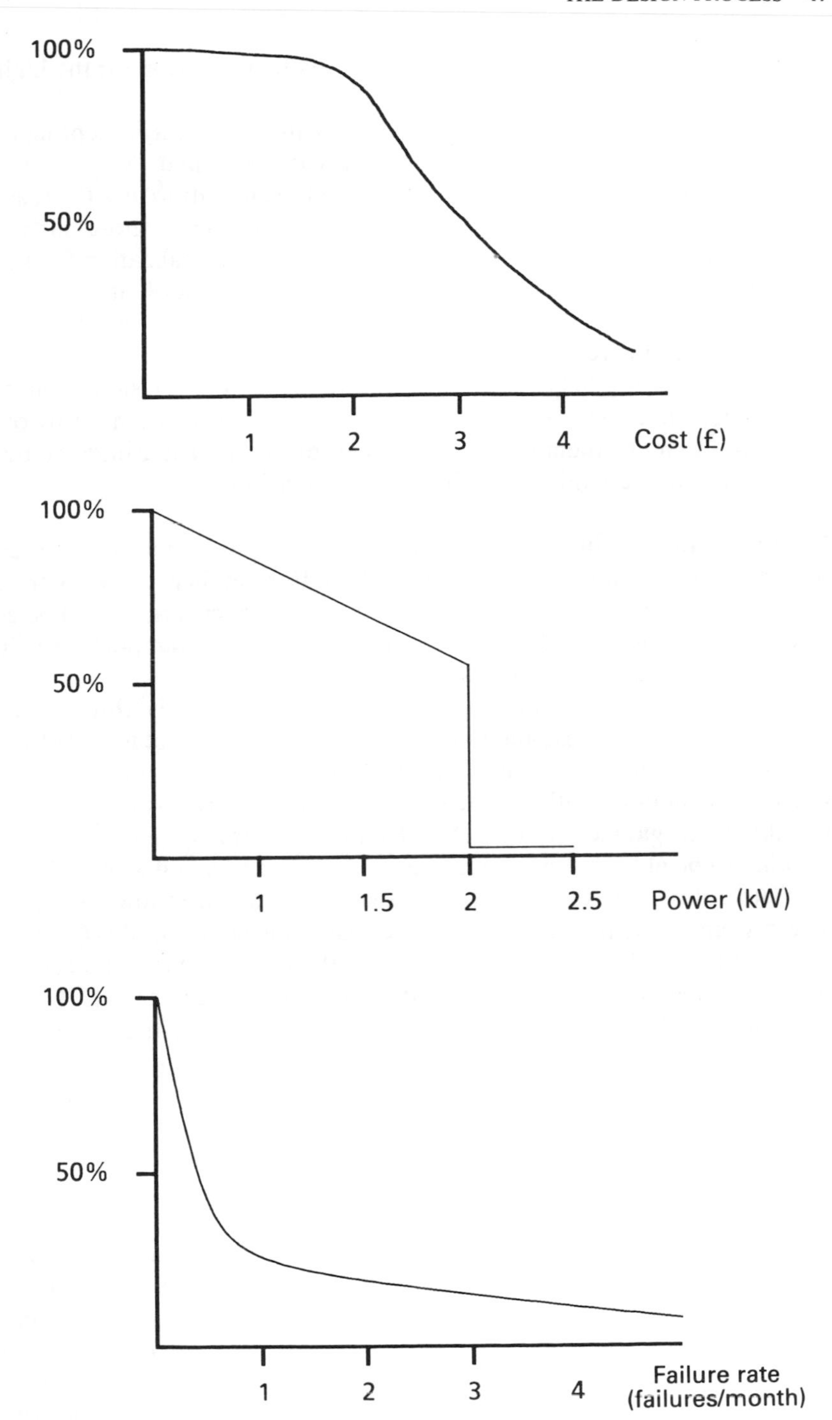

Figure 2.5 Utility curves.

6. *Making a choice.* A choice can now be made by looking for the highest score of the options.
7. *Applying a test of reasonableness.* This means not blindly accepting what the score line for the exercise gives. One can test in many ways. One can look at all the options that have been rejected and rationalize the reasons for rejections by looking at which of the criteria made those particular options score low. One can do the same sort of rationalization for all the options that scored high. Having done this one can look at some of the rejected options and ask whether the score could be increased by making changes in the areas of low scoring criteria.

   This analysis helps to confirm (or otherwise) the decision implied by the score line. This in turn either makes beneficial changes to other options to make them more attractive or brings a greater understanding and therefore confidence in the original decision.

Example of the Combinex method As a simple example, consider the selection of a new car using the Combinex method. For simplicity we will consider only three criteria. (Normally of course many more criteria would be considered, including some that relate to features, but the example here will be simplified for illustrative purposes).

Let us assume that the customer is considering only three criteria, namely cost, length of car and range in miles. We may speculate here that he is working to a limited budget, that he has a garage which is limited to a specific maximum length and/or he would like to leave as much space as possible in the garage for other uses. He may also have a specific reason for wanting to be able to drive a minimum distance on one tank of fuel.

The utility curves representing this set of criteria are shown in Fig. 2.6. The customer is 98 per cent happy if the cost of the car is £15 000 (nobody is ever 100 per cent happy about cost!). As the cost increases the customer becomes progressively less happy, until at £17 000 he becomes barely 60 per cent happy. The customer is perfectly happy with a car that is up to 4.3 metres long, but after that becomes progressively less happy. The requirement for range of the car on one tank of fuel does not fully satisfy the customer until it exceeds 700 miles. At any range less than that the customer's satisfaction is significantly reduced. These three curves (and there would normally be more) give a picture of the various satisfaction levels that the customer has relative to certain criteria.

Now let us look at the specifications of three different cars and how they match up to the requirements specified. They are cars A, B and C. The specifications for these models are shown in Fig. 2.7. The cost and length of each of the cars can be read directly from the specifications and entered on to the utility curves as shown. For range, this can be calculated using the figures for fuel consumption and the fuel tank capacity. The ranges for each of the cars can also then be entered on the utility curves.

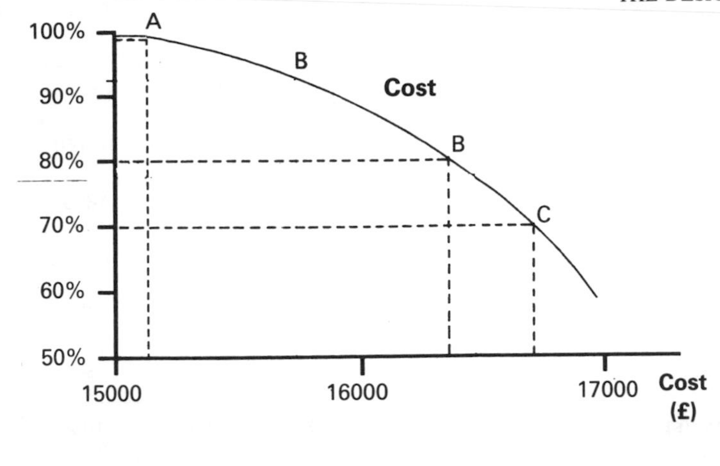

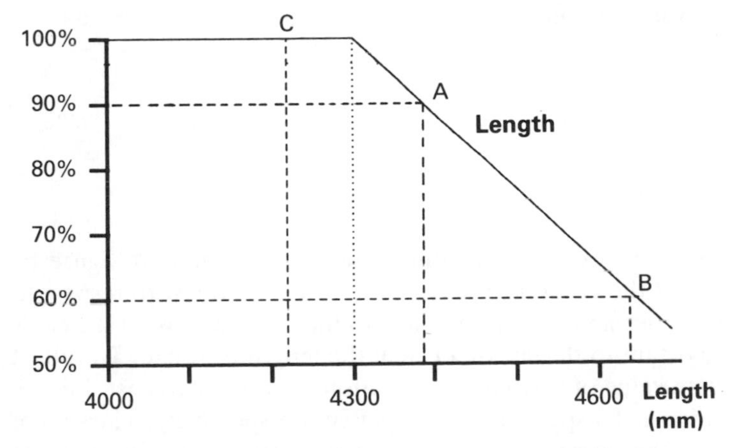

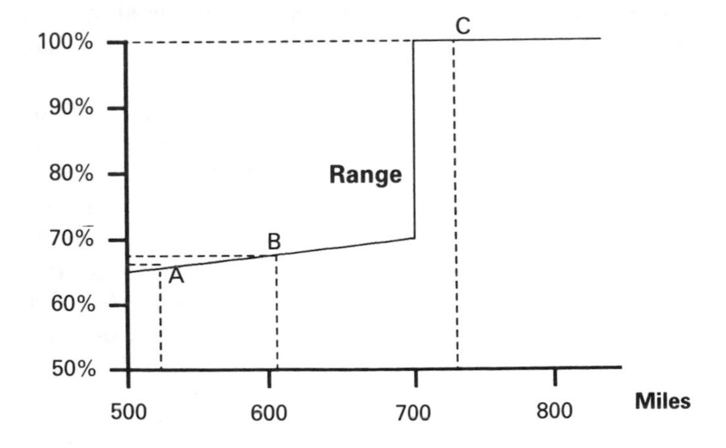

Figure 2.6 Utility curves for the selection of a car.

Feature	Car A	Car B	Car C
Cost (£)	15 182	16 300	16 750
Engine capacity (cc)	1991	2600	1800
Overall length (mm)	4385	4640	4227
Overall width (mm)	1645	1772	1670
Fuel capacity (litres)	62	75	80
0–60 mph in time (s)	10.6	9.8	10.2
Fuel consumption at 56 mph (miles/litre)	8.5	8.1	9.1
Range (miles)	527	605	730

Figure 2.7 Specification data.

The Combinex process now recommends that a weighting figure be assigned to each of the criteria so that their relative merits with respect to customer satisfaction can be accounted for. In this case the cost of the car is much more important to the customer than the length or range. To reflect this, a 50 per cent value of customer satisfaction is assigned to cost. Length and range are assigned 30 per cent and 20 per cent respectively. The sum of the customer satisfaction ratings should add up to 100 per cent. Then as a test of reasonableness, each individual rating can be compared with any other to check that these assessments are correct. In this case, is range about two fifths the importance of cost? Is car length 50 per cent more important than range? If these figures seem reasonable then they can be entered in the 'weight' column in the table of Fig 2.8. The score for each of the three cars can be entered in the 'score' column by reading the relevant satisfaction level off the utility curves. For example, car B at a cost of £16 300 gives a satisfaction level of 80 per cent, and car C at £16 750 gives only 70 per cent.

With all the scores entered in the table, the weighted score can be calculated by multiplying the absolute score by the weighting factor for that criterion. Thus car B with 80 per cent satisfaction on cost becomes $0.5 \times 80 = 40$ for the weighted score. Adding all the weighted scores for each car gives:

Car A 89.4
Car B 71.6
Car C 85.0

		Car A		Car B		Car C	
Criterion	**Weighting**	**Score**	**Weighted Score**	**Score**	**Weighted Score**	**Score**	**Weighted Score**
Cost	0.5	98	49	80	40	70	35
Length	0.3	90	27	60	18	100	30
Range	0.2	67	13.4	68	13.6	100	20
Total			**89.4**		**71.6**		**85**

Figure 2.8 Assessment table.

The highest score is achieved by car A and this is therefore the one to be selected on the criteria considered. However, as with the Pugh method, a test of reasonableness must be applied to the overall result and if two scores are close together the reasons for this must be carefully evaluated. For example, if car C was just a little cheaper and scored 80 per cent satisfaction on cost it would compete with A in the other criteria.

This example serves to illustrate the Combinex method. In practice, most selection processes are somewhat more complex, involving more criteria and often more options. In these cases this process proves invaluable to identify key differences in choice of options and also to clarify important but perhaps subtle attributes. The process requires good numerical data to maximize its usefulness.

2.3 BRAINSTORMING

In trying to generate new and innovative ideas, the team depends on the particular skills and personality of a small number of individuals who seem to have the ability to come up with clever ideas. Everyone else can be left feeling that since they do not seem to have this innovative flair they are surplus to requirements at this stage in the process. Nothing could be further from the truth. Everyone without exception has the ability to come up with new ideas.

Lateral thinking or thinking 'outside the box' is a useful technique for finding new ideas. Figure 2.9 is an example to test your skills at lateral thinking.

Figure 2.9 A problem in lateral thinking.

> Join the pattern of dots with four straight lines, without removing the pen from the paper once you have started, so that the line is continuous.

Lateral thinkers will think 'outside the box', and this is the clue. See Fig. 2.10 on page 54 for the solution.

Here is another good example of lateral thinking in a puzzle set by a few carefully composed sentences. In this case the solver must weigh each and every word of the poser, and apply lateral thinking to find the solution.

> A man enters a field carrying a small parcel. In the morning he is found dead and the parcel is unopened. How did he die?

The words give all the clues necessary to resolve the puzzle. The first clue is given in the phrase 'enters a field'. Most people will visualize this as a man climbing over a gate or hedge, or even on some vehicle such as a tractor or a bicycle. But there are other ways of entering! Secondly, there is the 'small parcel'. Visualize the various ways in which he could be carrying this—under his arm, in a bag, on his back?

By using lateral thinking one could determine that the man could 'enter' the field from below or above, which opens up a new range of possibilities. Lateral thinking may also provide the possibility that the 'parcel' may have been carried on his back, like a haversack or a parachute. The new possibilities quickly open up the plausible solution that he fell from a plane and his parachute did not open, causing him to be killed in the said field. This is another example of how by thinking outside the box, or laterally, a new range of solutions can suddenly be made available.

Even if you do not think of yourself as a lateral thinker, it does not mean that you cannot have creative thoughts. As a member of a team there are group methods that can be used to bring out the best in innovative thinking. One of the most productive is *brainstorming*.

The success of brainstorming depends on allowing the interactions within the group to trigger new thoughts in each individual's mind. We are often aware of this happening in everyday life where something someone says leads us to recall or develop a mental picture of a particular event or subject. Sometimes this will cause two people to have a similar thought as a result of hearing the same statement. In brainstorming there needs to be a strong discipline, otherwise it will not work effectively. Therefore rules are laid down before the session starts:

1. No criticism
2. No evaluation
3. One at a time
4. Write everything down on a flipchart or wallboard
5. Record the contributor's words faithfully—do not summarize or interpret
6. Work quickly.

A typical brainstorming session should be conducted as follows. The group assembles and agrees on the subject for brainstorming. This is then written down by the facilitator so that everyone can see it throughout the session. The statement should be clear and in the form of a subject statement that reflects the problem to be solved. A person is appointed to write down the ideas and a flip chart or display board is used to do this task. It is important to write down everything that is suggested and that every member of the group has visual access to the list as the brainstorming proceeds.

When everyone is aware of the rules and has concurred the subject statement, then the brainstorming can begin. Suggestions are called out either in turn (that is going around the room) or at random. At first it is sometimes better to go around the room in sequence until the ideas become less prolific, and then to revert to a random approach. Every idea, however silly or obscure, is written down on the board. No idea is rejected because it is impossible to tell which idea, however remote it appears from the subject, will trigger the next thought. Indeed, it is often a good idea to throw in a crazy idea as the flow becomes sparse just to change the trend. Often one can see this rekindle the prolific flow of suggestions. Some triggering techniques are to bring in an animal ('Lift it using an elephant's trunk!') or use nature as an example of achieving the solution.

Once a full list has been generated and the team's input seems exhausted (usually when there have been no new suggestions for two or three minutes), the team should look at the overall list to see if there are any trends emerging. For instance, how can some ideas be grouped together as similar themes? Also at this stage all ridiculous ideas are rejcted (elephants' trunks, etc.), as well as any that are obviously outside the realms of adoption (for cost reasons, for example).

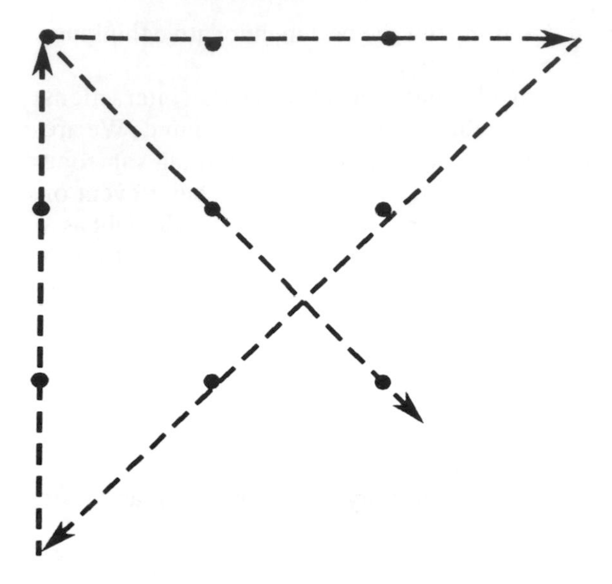

Figure 2.10 Solution to lateral thinking.

Another useful technique at this stage is to display the grouped ideas into a 'fishbone' chart (Fig. 2.11). This has the effect of graphically assembling the ideas into some semblance of order and may actually generate a new idea or enable the combination of ideas. There are many ways in which the list can be manipulated and refined and the process from this point is dependent on the problem to be solved and the way the team prefer to handle it.

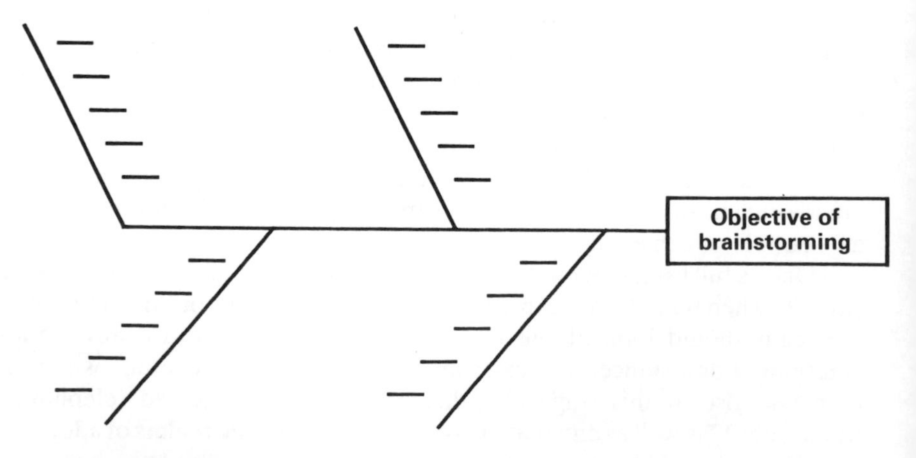

Figure 2.11 A fishbone chart.

2.4 RELIABILITY GROWTH

When products are initially conceived, companies must have an outlook for the ultimate reliability required in the product based on the known customer requirements for the product. This outlook is usually obtained from either historical data based on products that went before, on competitive products whose performance is known or on a projection based on knowledge about how various functions will perform and their various failure modes in the field. If the reliability of a product is to be kept under control during the time that the product is being developed then throughout the process of product design and development the reliability achievements must be monitored and compared with what is expected at the particular stage of the development. Historical data on reliability growth normally held within the company is often used to give an indication of how rapidly reliability of a product can be improved as it goes through the various development cycles. What is often found here is that the rate of growth follows a fairly set profile for a particular industry and product. This means that without some revolutionary change in the process, the team's progress on reliability improvement is fairly predictable through the phases. Once this profile is known from the data associated with previous products, it can be used reliably as a planned growth profile until there is some significant change in the process to cause a shift.

Achieving even this historical profile of reliability growth demands a careful understanding of how to effect improvements in performance, and this in turn demands an understanding of the nature of failures and errors that occur during a design. An analysis of the growth in reliability and the elements that make up the full set of failures in the various phases is helpful to this understanding.

Every product has a failure rate. Some of course are better than others. Refrigerators and TV sets, for instance, now have very low failure rates whereas copying machines and washing machines are not so reliable. Every product strives to 'grow' reliability over the years by the use of new materials, new technologies or better design or manufacturing processes. In order to understand better the ways in which reliability of a product can grow, let us start by looking at the reliability of a product at the point when it is launched. Reliability is often assessed by measuring the inverse of reliability, namely failure rate, and this can be measured in terms of failures per 10^6 cycles of the product's operation.

Consider the reliability of a product at launch. It will have a failure rate of x failures $/10^6$ cycles or operations. This rate will comprise failures which can be categorized into one of three groups (see Fig. 2.12):

1. Those failures arising from manufacturing defects (*manufacturing failures*). These are due to some failure during the manufacturing process

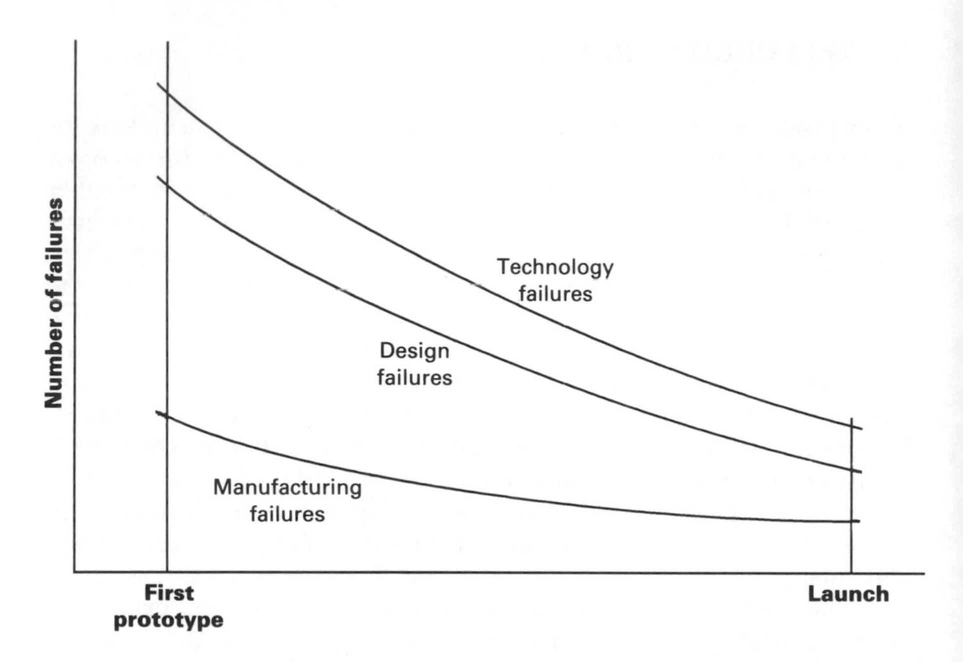

Figure 2.12 Failure profile during development.

to meet the laid-down specification for parts manufacture or the assembly process. This is usually a failure attributable to an error in the manufacturing process itself, such as a setting, an adjustment or a human error in the assembly of parts. Examples would include the failure to tighten a screw to its specified torque or the omission of a part or setting.

It is important that the designer appreciates that responsibility is shared for these failures, for it is the design itself that allows the shortfall to occur and should not be considered solely the fault of the production operator.

2. Those failures arising from design faults (*design failures*). These are failures that occur due to the failure to follow design principles. The design principles referred to here are those principles that state the constraints of applying a particular design embodiment. They are based on years of experience and analytical study and are usually found in design manuals or guidelines issued by the materials or component manufacturers. They represent a failure of the designer to apply known constraints in the design such as the correct selection of material, adequate allowance for stress, etc. In general they can be resolved immediately by correcting the error, use of the correct material, applying adequate stress relief radii, etc.
3. Those failures arising from technology faults (*technology failures*). These are failures that occur as a result of the system or subsystem trying

to operate outside the latitude for which it was originally designed. Typical examples of this occur when the environmental conditions exceed those anticipated during design. Conditions such as excess temperature or humidity may occur during operation of the product which were not specified or accounted for during the design. More often a design that is marginal on latitude may as a result of a tolerance build-up between a variety of components begin to exceed the limits of its operational capability. These failures cannot usually be corrected easily by making a simple change to the design. They require a significant redesign to enable a broader latitude to be achieved and are the most serious failures a designer can be faced with.

Assuming that the design contains sufficient manufacturing input to ensure that manufacturing failures can be contained to a reasonably low level by addressing manufacturing techniques and quality after the fact of design, let us consider the other two elements. Given the generally accepted rule that the reduction profile of failures (reliability growth) during the development process is predictable for a particular industry making a particular product, we can study the make-up of reliability at any point prior to launch.

Let us now consider the period prior to launch. During this period the failure rate is dominated by two of the three failure types, design failures and technology failures. As I said earlier, the improvement in the reliability growth through the development process is fairly predictable for a particular industry making a particular product and so we can project back from launch the slope of the reliability growth curve. At any time during this reliability growth period, the failures making up the total failure rate will consist of both technology and design failures. The combination of these failures will be represented by a ratio whose value will be dependent on the particular point in time of programme development.

Let us look at the improvement rate of these two categories (Fig. 2.13).

The level of design failures can continue to be improved as the design process proceeds. A failure due to a bad choice of material can fairly easily be rectified by changing to the correct choice of material. Similarly, a design error in not using a radius to eliminate a stress point can easily be remedied by adding the appropriate radius. Here the only difficulty is finding out where the design failures have occurred. This is traditionally done by testing, but there are other methods that enable the designer to identify shortfalls in design before hardware is built and tested and these are dealt with later (see Sec. 3.7).

At the point where the production intent design is established, that is at the end of the concept phase, it is essential that the proportion of technology failures is low compared with that of design failures. This is because at this point all technologies will have been chosen and the architecture will have

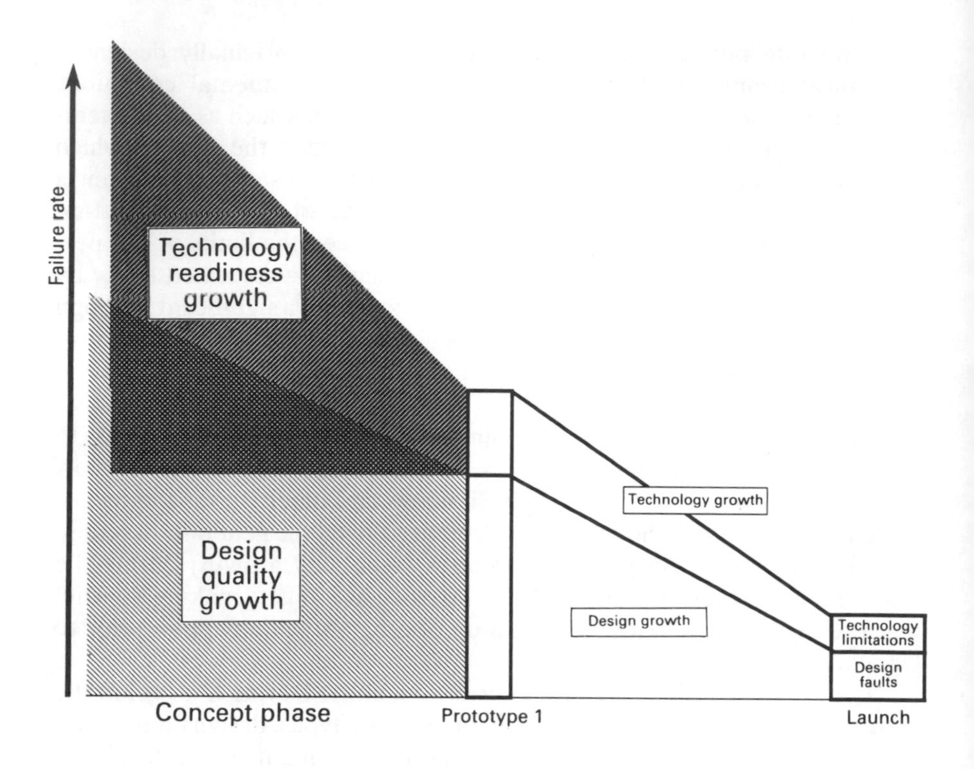

Figure 2.13 Reliability growth through design.

been defined and if the number of technology failures is not low there will be little opportunity after this point in the process to reduce it. Using an extreme example, if the propulsion system for an aircraft is chosen as propeller driven and this technology cannot achieve the speed that is required, then the propulsion system must be uprated to a jet engine. It is obvious that this is a major disruption to the design, not only in terms of the type of engine but also in terms of the system (for example fuel, controls) that are required to support the new engine. If a technology has to be changed after the point when the design has been started, major disruptions to the integration of the design are therefore likely. The reason for this is that during the concept phase all the work will have been done to establish the compatibility and integration of all the technologies. Any change to any one of these is highly likely to have a serious impact somewhere else in the design. Additionally, the fact that space and configuration of the various elements have been allocated means that any change of technology will probably mean a change of architecture and this will have a significant knock-on effect to the design.

Therefore if mature reliability is to be achieved by the time the product is launched, the level of technology failures must be at a low level prior to the start of the design. This state is called technology readiness.

The Concept of Technology Growth

A technology can be regarded as an entity which grows in maturity and stature as more and more work is done on it and as more understanding is gained in it. When selecting a technology for a particular function in a design, it is important that the designer has a full understanding of the growth history and the growth potential of this technology.

Let us trace the growth of a technology in a generic way to help to understand this concept. At some point a technology is born as a result of an idea which solves a problem or takes engineering a step onwards from the conventional. The birth of a new technology is an invention. The jet engine, the hovercraft, the movie camera are dramatic examples of these which most people marvel at. In product design, however, the invention of a new device for motive power, or creating images or simply introducing a new material may be regarded as a new technology. However dramatic or mundane the new invention may be it will have limitations in its capability and until these limitations are understood and defined the new technology cannot be used successfully in a product.

For example, the electric car has been in the headlines for many years now but has not yet become a viable product. This is directly the result of unreadiness in the technology itself, because the environmental pressures have never been stronger nor the competitive advantage more attractive. But until the technology of electrically driven power for the motor car can meet the stringent demands of the existing technology, the internal combustion engine, the transition will not occur. In a similar way, solar energy awaits its own technology readiness. Although it is seen in some specialist applications it is not yet ready to compete with more conventional sources of energy.

There are always tremendous pressures to extend a technology outside the boundaries known to be those within which it currently works. Knowing the boundaries within which the technology must work produces enormous pressures to extend the capability of the technology outside these constraints. For this to be achieved the technology must be developed, refined and stretched until it can be extended to new realms of performance. Sometimes this will take many steps along the way, and at each step the designer must understand the new limitations for the technology. At some stage the technology will reach a point where it cannot be extended further and an alternative to the technology will be sought. The cycle starts again as invention makes that leap across to a new technology, extending capabilities even further.

In design it is important not to make the leap to a new technology for the wrong reasons. It should only be made when existing technologies have been shown to be incapable of being extended further. As we discuss technology readiness in the forthcoming chapters, we will see how an important estimate can be made of how close a technology is to becoming fully extended.

2.5 TECHNOLOGY READINESS

The point in the programme where the design is embarked upon is the point where there has to be a commitment to spend a significant amount of money. After this point the level of manpower in the team will build up, there will be a need to build a number of engineering models for testing and the time will draw close where substantial amounts of money will be required for production tooling. It is therefore a key point in the product programme and management will need to be absolutely sure that all decisions that have been made carry an acceptable risk level and that the technical status is looking good. Technically at this point the design must have minimal (if any) technology or architectural problems associated with it. That is to say, each of the technologies chosen and the set of technologies as a whole must not be at risk in terms of functional performance or integration. The technical work in the concept phase must have established that this situation obtains prior to commencing the design phase. Bitter experience has shown that failure to do this may at a later stage force the programme to change a technology or the architectural configuration of the product and this can be catastrophic. At best it will delay the programme for many months. The basic reason is that no one function can be considered in isolation. If the technology to achieve that function is changed or even its relationship with other technologies is changed, then the knock-on effect can be enormous. Just a simple example of where an element of a machine is underpowered and has to be substituted by a larger unit clearly causes an immediate space problem. This may require the movement of other subsystems, which may necessitate a change to the framework of the machine, etc. Usually changes of this kind are much more complex and serious.

In order to be sure that changes in the design during the design phase will be able to be accommodated it is essential to establish that all technology failure modes have been satisfactorily prevented or resolved. Once this stage is reached for a particular technology, that technology is said to be ready, and once the total system has reached this stage technology readiness has been achieved.

How, then, can we be sure at the end of the concept phase that the technical status is advanced enough for the programme to have reached technology readiness? Maurice F. Holmes defined five criteria for technology readiness:

1. The critical parameters of the subsystems and the system must be defined.
2. The failure modes of the subsystems and system must be defined.
3. The latitude of each subsystem and the system must be defined.
4. Manufacturability must have been assessed and found acceptable.
5. The system hardware must have been demonstrated.

When these five criteria have been met the design is considered ready to move to the design phase. Let us look at each of these criteria in greater detail.

1. *Critical parameters.* All critical parameters, that is those parameters that have a primary effect on the function of the system and therefore have a significant effect on the failure modes, must have been defined and nominal values specified using either analytical or empirical methods. This is the starting point of critical parameter management (see Sec. 3.5). These parameters will be the language by which the design is described and will serve to define the design intent, validate the drawings or design files and audit the hardware in all phases of the programme. Since they are the key definition of the design, it is essential that the list is complete and desirable that both nominal values and their range of tolerance are specified at this stage.
2. *Failure modes.* Subsystem and system level failure modes must have been identified for the selected technologies, both individually and as a set. Doing this helps to create a full picture of the operation of the system and is invaluable in improving the reliability of the product.
3. *Latitude.* The safe operating latitude and its sensitivity to each of the failure modes and the set of critical parameters must be understood. Knowledge of latitude, whether wide or narrow, will be valuable knowledge during the optimization process and also while trying to reduce costs and ensuring good manufacturability of the parts and assemblies.

The above three criteria are essential for technology readiness. The two criteria following are desirable but are not an essential part of the technology readiness assessment.

4. *Manufacturability.* A study should have been conducted to make sure that there are no parameters in the design that demand a level of manufacturing accuracy or process capability that is not available. If this study highlights one or more areas where a new manufacturing technology is required or where standards higher than the norm are required, the team must pursue actions to find solutions to these problems. If solutions cannot be found, of course there is no point in going ahead with a demand on manufacturing that cannot be achieved. However, if the capability required can be achieved, say, by making more cost available, perhaps by reducing yield requirements in production, then of course the situation

must be weighed as part of the business case and a programme decision taken accordingly.

It is not necessary at this stage to have full vendor commitment to the manufacturability of the product, unless there is a manufacturing difficulty, the risk for which is dependent on the vendor's capability of carrying out the manufacturing operation.

5. *Demonstration of hardware.* Ideally, the demonstration of functional performance of the system is required to achieve tcchnology readiness, and this should be pursued as part of the concept phase if the programme has the capability in terms of cost and time to do so. However, there are other ways of 'demonstrating' function without the need to build expensive hardware and this should be considered as an alternative option. Simulation, modelling and analysis can be employed to establish the robustness of the system against noise factors (that is parameters outside our control) and this may be a much more cost effective way of satisfying this criterion. Further, it is important to identify any new interactive effects of the system (or show that none exist), including interactive effects on latitude or failure modes.

The importance of achieving technology readiness cannot be stressed too much. A product programme that goes ahead without technology readiness is doomed to failure. More often than not, of course, failure of a programme occurs because technology readiness was never evaluated rather than lack of it ignored. When a programme is committed to a production intent design the confirmation of technology readiness is essential.

The Rolls-Royce RB 211 engine was originally designed using carbon fibre fan blades. The technology for these blades was not ready at that time and they subsequently failed mandatory certification tests. The only recourse available was to embark upon a major redesign of the engine which would use conventional blades. The engine had to be significantly bigger than originally intended and this delayed the programme so much that orders for the engine were significantly reduced, contributing to the eventual bankruptcy of the original company.

2.6 THE IMPORTANCE OF LATITUDE

Why does a design need to have latitude? One can understand the importance of failure modes and the need to understand them in order to provide a more robust design. It is also obvious that the parameters that control the function must be in the control of the designer wherever possible and for those that are not they must be capable of being tolerated by the overall design. Latitude seems like a luxury that is not essential for good design quality. In a way, latitude can be thought of as unimportant when one is thinking of a single piece of hardware operating according to the functions as prescribed by the design. However, as soon as the design is thought of in the context of the production of a number of units then latitude takes on a different role. Firstly, it is important to understand that no parameter has one absolute value. Every parameter has a nominal value and a range over which this nominal value can vary. Correctly worked out this range is what can be tolerated in order to satisfy the function in which it plays a part, and is then termed the tolerance of this dimension or parameter. There is no exception to this fact throughout all the parameters of the design, and the only way in which there is variation is in the narrowness or broadness of the range. Therefore, when one is producing parts for a product in quantity, every parameter on every part will have a slightly different value to that of its neighbour.

A good designer will have worked out what the tolerable range for each of the parameters is in order to ensure that the assembly functions correctly once it is completed. The designer will want to make sure that these tolerances are as wide as possible to make the part easier to manufacture, but also to be sure that even when a parameter is positioned away from its nominal value the part will still function as required. In manufacturing the engineers will have arrived at a process that gives them the best opportunity of keeping all of the parameters as close as possible to nominal.

Now let us consider the progress through the design of the product and the development of the manufacturing processes as the programme goes from its initial design stage to final manufacturing for production. The design engineer is striving all the time to design the parts of the product so that their tolerance bands and latitudes are as broad as possible. Engineers want to increase the bands continually because they know that the more they do that, the easier the parts of the product will be to manufacture and the less likelihood there is of failure due to responses of functions going outside their range of latitude. This translates into a cheaper and also more robust product.

On the other hand, the manufacturing engineer is trying to arrange the production processes so that they have the minimum variation as each part is made one after the other. This is done so that parts are delivered that are more likely to be within the ranges that the designer requires and will therefore contribute to a more reliable performance once all the parts have been assembled together to form the final product.

Often at the start of a programme the designer's latitude range will be close to or even narrower than the variance that the manufacturing process can achieve. This of course is an unacceptable state of affairs in the final product and would result in the product being very susceptible to failure. Therefore the design engineer continues to broaden latitude while the manufacturing engineer tries to narrow the variances in the manufacturing process (Fig. 2.14).

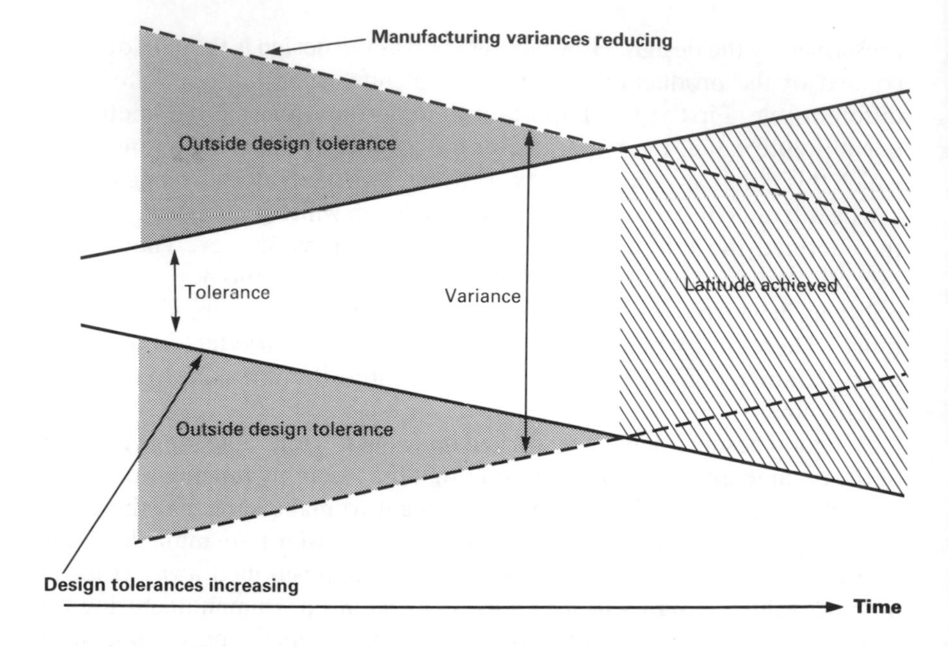

Figure 2.14 The importance of latitude.

It is obvious from Fig. 2.14 that until the manufacturing variances are within the range of design tolerances there is no certainty the design will function correctly. Of course, in the early stages of development when manufacturing variances are wide and design tolerances narrow, *some* of the parts may be satisfactory. These will be those parts that fall into the band of narrow tolerance defined by the design. However, this situation does not provide robustness in the product and hence cannot be allowed to prevail. Quite often companies think of robustness as being the responsibility only of manufacturing in providing parts well within design tolerances. This concept of latitude shows clearly how the responsibility must be shared and how the designer should consider very seriously the tolerance ranges which are defined and not just state tolerances that are thought to be those achievable by manufacturing.

Therefore, for the designer, the concept of latitude is of great importance. The designer must be concerned not only with the latitude as it occurs in the baseline (or time–zero) design but also understand and accommodate other areas where latitude can be reduced. Primarily one can think of latitude reducing as wear takes place in the product or when replacement parts have to be fitted. The testing of a product must also be organized so that the actual latitude of the functional response is measured. If we cannot measure the latitude, how can we confirm that the latitude that we thought we had in the design actually was achieved?

2.7 THE PROCESS OF DESIGN

In general, today, the design process consists of a series of iterations which are necessary to achieve the ultimate goal of a high quality design that satisfies the requirements of the customer. The number of iterations and the form that they take are dependent largely upon the complexity of the system under design and the level of design expertise and design support available to the team. For example, a team with a large amount of analytical support and expertise may be able to minimize the number of iterations required simply by being able to either model or simulate the operation of the design by creating mathematical or analogue representations. It is a fallacy that modelling requires extensive skills or heavy computer capability. Many engineering problems can be evaluated analytically using quite simple modelling methods. Many designers will use cardboard replicas to gain a better understanding of perhaps the kinematics of a mechanism. (Many CAD systems are capable of this also.) Simple mathematical calculations can enhance knowledge enormously without more than a fairly simple use of algebra. Others with a more designer oriented team may rely more on what has gone before and need to create test models with which to evaluate performance and reliability and then put effort into resolving the failures and problems that come out of the testing.

The precise formula for the development of the design of a product can therefore vary enormously. It is, however, an accepted fact in most modern design communities that the iterative process is very costly, both in terms of time and money, and that any means towards reducing the numbers of iterations is well worth pursuing.

Typically the stages for developing the design of a product are arranged to evaluate the following questions:

- Is the proposed design feasible in terms of its operation and function?
- How can the system be integrated?
- What are the engineering problems associated with the proposed design?
- What are the manufacturing, performance and reliability problems associated with the proposed design?
- What are the problems associated with the hard tooled parts and the assembly tooling for the product?

Examples of the types of hardware and design iterations that are required to answer these questions are listed below.

Feasibility Rig

This is a rig originally designed to provide the first embodiment of the design and an understanding of the shortfalls in the fundamental thinking of the functional operation of the product. For example, a designer may be relying on a particular new technology or invention to provide a function critical to the performance of the product. Embodiment of this into a feasibility rig will provide confidence that the original thinking for the design function was correct or alternatively has certain identifiable shortfalls. The need for a feasibility rig is reduced if the design is an evolution of a previous product with little or no new technology or invention in it.

It becomes clear that if one is seeking to reduce the number of iterations by eliminating a feasibility model, then one should either be seeking to minimize the use of new technology or discovery in the design or alternatively should try to find a more effective way of proving feasibility other than by the use of hardware and testing.

Integrated Rig

One of the most important aspects of arriving at the final production design at the earliest possible stage is to embark on the 'production intent' design as soon as possible. The production intent design is the design that, albeit immature and incomplete, is the most sincere attempt that could be made via the chosen production processes. For example, if a component is to be made by moulding in plastic, then in the production intent design this part is shown designed as a plastic moulding and not as, say, a part machined from solid. In the build of the rig it may well be made by machining solid raw material, but the part would be as close as possible to its shape and appearance in production. Strength and function must be assessed separately. Another important attribute of the production intent design is that it should reflect the final relationships (as best known) of parts and assemblies. In other words, it is the best current estimate of the final design that the designer can accomplish.

The integrated rig reflects the production design intent. It is designed and built to establish the answers to such questions as 'What are the best relationships between subsystems and parts?', 'What space constraints are there?', 'Are any new problems created when functions have to work together?' In terms of performance and testing the integrated rig would be the first embodiment of the complete system hardware and would be used as a vehicle, during build, to identify and resolve any physical integration problems and, during testing, to identify and resolve any functional interaction problems.

Again it is obvious that the more work that can be done to resolve these problems before the building of what will be an expensive rig, the more effectively the design process will proceed. For example, the very process of testing for interactive effects in a system can be a long-term process, whereas modelling techniques using widely available computer software can be equally effective and significantly reduce the design, building and testing effort required to collect the information by empirical means.

Engineering Model/Prototype

The terms engineering model and prototype are mostly interchangeable, depending on the nomenclature of the particular industry. They are models required during the design and development process to evaluate the functional operation of the system. They are, almost by definition, required to be of production intent design. They are the early embodiment of the production intent design, required to see how the total system works as a completely integrated whole. They are used to evaluate performance, establish reliability and unearth problems that arise from the engineering of the product. This is normally a complex and time consuming activity and involves long test runs to establish statistical data and life effects and to pinpoint precisely the root causes of problems. Once this has been established initially, the same hardware is used to evaluate the best solutions to the identified problems and then to test the effectiveness of these 'fixes' as a second round of performance testing.

It is often the case that more than one generation of this level of test hardware is required to bring the reliability and performance of the product up to the required levels. Consequently, the cost and schedule load to the programme is one of the heaviest at this stage, and there is therefore a great incentive to minimize this in some way.

Preproduction Models

Preproduction hardware, or pilot hardware as it is sometimes called, serves the same purpose as the engineering models, in that it strives to establish and raise the level of the functional performance of the system. The difference here is that the preproduction hardware is also required to test and prove out the hard-tooled components used in these models and the assembly tooling used to put them together. In other words, it is a 'dry-run' of the production process using the proper parts and the production methods of assembling them. This, as all product engineers will know, is a major phase of the process of designing and developing a product. It brings together the designer, the engineer, the production engineer, the assembly worker and the quality control office. It is hampered in some ways by the desire to minimize change and this creates a well-known engineering dilemma. The dilemma is that unless changes are made to the design, the reliability and performance of the product will not be changed in any way. However, every change that is made costs large amounts of money and time in changing expensive tooling in order to convert the parts. Thus, in striving to reduce delays and reduce the costs of change, the programme must minimize the number of changes but also ensure that those changes that are needed to improve reliability are implemented. Therefore, once again, there is a strong desire to minimize the extent of this phase of the activity.

Figure 2.15 shows a typical track of hardware iteration for a consumer

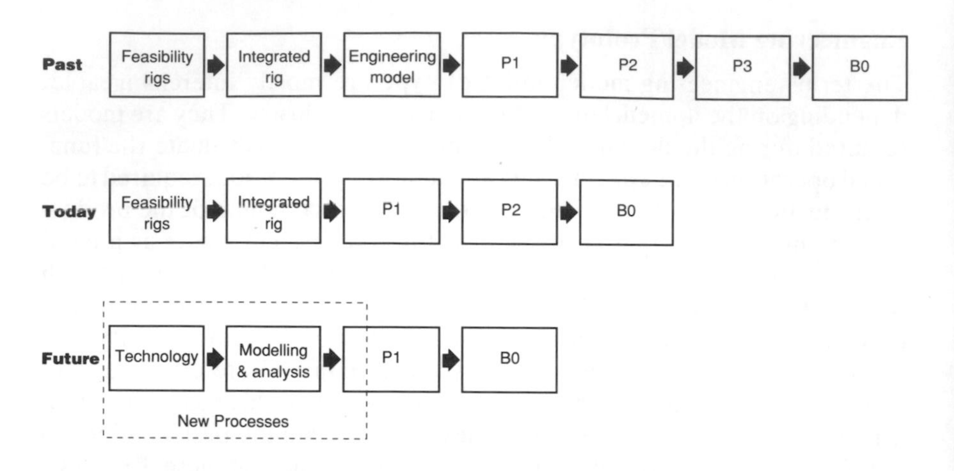

Figure 2.15 Hardware iteration.

product and shows how improvements have been made over recent years to reduce the number of iterations and consequently the overall schedule. In general, the most progress has been made at the prototype and preproduction end of the scale, where there has been a growing understanding of the need to practise concurrent engineering and work together with manufacturing and parts suppliers at an early stage in the design in order to minimize the problems that are usually encountered during the transition from engineering into manufacturing. As has already been said, this has been achieved by getting the engineering personnel and the manufacturing personnel together at an early stage in the process. This has ensured that the design has had more attention paid to manufacturability problems and has enabled new design and engineering techniques and new manufacturing methods to march together rather than being out of step with one another.

Additionally, there has been a greater awareness of the need to start early with a production intent design, and this has tended to reduce the necessity of reiteration of the engineering design due to a need to accommodate particular manufacturing limitations and constraints. These have a serious knock-on effect, and until the production design intent is addressed, the whole of the system design is in jeopardy, as any single need for change may create a major system integration problem.

The long-term target for most industries in terms of numbers of iterations of hardware is likely to be one prototype and one preproduction model before manufacturing begins. It is foreseeable that with continued application of concurrent engineering (including any outside suppliers of parts), the single step transition from prototype to manufacturing could be achieved. To reduce the numbers of development and engineering models requires a new approach. There is a requirement to readdress the way engineering deals with the introduction of new technologies or discoveries and the way the fundamental design process is conducted if progress is to be made in this area.

Referring back to Sec. 1.3, design has three attributes. Engineers and designers are often heard to express these when discussing a design. For example, a design exists primarily to deliver a function. A lever is a design to provide mechanical advantage. A spring is a design to deliver force or torque. However, designs also have two other fundamental attributes, form and fit. Form is required to make them aesthetically pleasing or acceptable and fit is to ensure that they integrate with their surroundings or neighbouring designs. Thus form, fit and function are the three attributes of design. How do we know whether a design meets the requirements for form, fit and function? Generally a check on whether a requirement is met or not is carried out by measurement.

In the case of form, measurement in the generally accepted sense of the word is difficult. However, by setting standards for form or providing baseline examples for comparison, a satisfactory means of measurement can be devised. Certain aspects of form, such as straightness or texture, can also be measured directly by using the correct instruments.

The confirmation of fit is much easier. Dimensions related to fit can be easily measured using conventional measuring tools and compared with dimensions and tolerances clearly laid down in the design, usually through geometric tolerancing on the drawing.

Function is much more difficult to measure directly. It is usually evaluated by some kind of testing. If we are to evaluate function on the same level as form and fit, we need to know a great deal more about what it is that makes a function operate effectively and reliably. Looking at what goes in to function to make it occur and what is its output when it operates effectively is one way of providing this understanding.

Essentially all functions can be described by using a noun–verb phrase, such as valve opened, torque delivered, lever turned, etc. A study of what inputs need to be in place to cause the function reveals that there are two types of input. There are those that are under the control of the designer and those that are outside this control. In order to distinguish between these two, the elements that the designer can control are called control factors and those that are outside the designer's control are called noise factors.

2.8 ENVIRONMENTAL DESIGN

Today's world is increasingly concerned with our environment, and rightly so. The continuous production of products for an ever-insatiable world is becoming a serious problem which will need years of careful attention, firstly to stem the tide of pollution and secondly to provide controls that will protect the environment in the future. We are all familiar with the dangers of oil pollution and the causes of global warming, but the lobby of opinion for much less obvious pollution is yet to be heard at full volume. Nevertheless, there is growing concern over things like acoustic noise, the use of energy and the disposal of parts and products that have come to the end of their useful life. By the time this is read it may be commonplace to be returning packaging and spent containers to the place where they were bought and the cost of disposing of an appliance may have been paid for in the purchase price or may be a significant cost of ownership of the product. It may not be long before governments and other users alike will not be satisfied with a product that is not either recyclable or made of some recycled parts. The effect that all this will have on the designer is significant.

Already we see products advertised that are deemed to be recyclable or environmentally friendly. The design of these products has been very carefully thought out in order to be able to make the changes to support such claims. More and more the complex business of recyclability and energy usage will need to be addressed as a key part of the design.

For example, it is now quite possible and economically sound to reclaim some parts of products and reuse them in the next or new versions of the product. Electrical components, for instance, can be reclaimed, tested and verified and fed back into the flow of 'new' parts for a product. There is already a recycling process for plastic materials which will enable parts made of certain types of plastic to be ground up and recycled into the plastics industry for use in lower grade products. Any material can be a potential candidate for recycling. Furthermore, materials that require disposal, due to their toxicity or unsuitability, will be expensive to dispose of properly and there will be harsh penalties for illegal disposal.

The design of products should take into account three areas of opportunity to help to protect the environment.

1. *Design for reuse.* Here consideration may be given to the type of materials and components used so that a part or assembly may be able to be cleaned up or refurbished to be reused in another similar product. Design for reuse also includes the selection of materials that have themselves been reclaimed. In addition, careful design for robustness may enable a part to last more than the lifetime of the original product, enabling it to be reused in another product.
2. *Design for recyclability.* This involves mainly the selection of the type of materials that can be reclaimed and then used as a raw material in some later product. Design guidelines have to be laid down for this. For example, a paper label on a plastic part will render the part much less able to be reclaimed because the paper will contaminate the purity of the plastic. The designer should mould the label into the plastic instead.
3. *Design for easy and safe disposal.* Here designers need to be encouraged to use materials that are easily and safely degraded. There are some

misconceptions in this area. It is said that the scientists investigating the effectiveness of land-fill sites date the layers by reading the dates on the newspapers! Therefore our belief that paper and cardboard are readily biodegradable may not be true after all.

The designer of a product needs to be aware of these new aspects of environmental requirements and to take account of them in the design.

CHAPTER

THREE

DESIGN FOR MANUFACTURE

3.1 THE QUALITY LEVER

The quality lever (Fig. 3.1) can be used to depict the relative effects of the various tools and methodologies at our disposal. Some of the processes that we can use have a major effect when used early in the programme while others represent reactive activities that occur much later. The quality lever is a model showing how the 'mechanical advantage' of some of the tools can be applied.

Quality function deployment is a tool that delivers a major leverage to the quality of the product. It represents the vertical axis of the Kano model. In other words, it is a major thrust for delivering customer satisfaction and can be applied early in the process as well as continuously throughout the

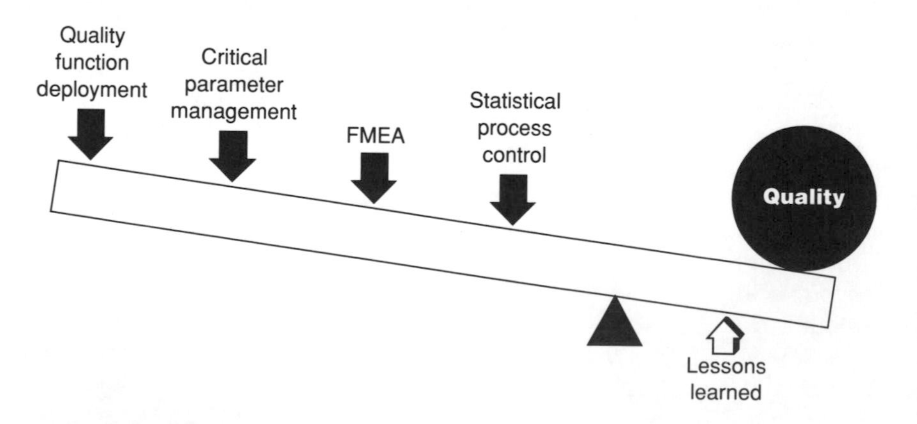

Figure 3.1 The quality lever.

process. Next comes critical parameter management which gives a firm understanding of the functional capabilities of the design. This can also be applied throughout the product development programme.

Failure modes and effects analysis (FMEA) is a tool that anticipates potential problems and tries to put in place measures that will eliminate or detect the problem. FMEA can be applied both for the design element of the programme as well as for the later manufacturing processes.

Closer to the launch of the product, statistical process control is another positive force for improving the ultimate quality of the product. Once a product has been launched, any activities post-launch will inevitably be retroactive. That is to say, they will be activities that tend to fix problems that have already occurred, rather than activities of prevention.

In the succeeding pages of this book a number of tools and methods will be described. Some will have greater ‘mechanical advantage’ than others. All will be a positive contribution to improving the quality of the design and ultimately the product.

3.2 QUALITY FUNCTION DEPLOYMENT

Quality function deployment (QFD) is one of the many design management tools that originates from Japan. It is essentially a method by which a great deal of information about a particular project or design can be assimilated on one page and one chart to enable the designer to make important comparisons and have a greater understanding of the total system. The quality function deployment chart will bring together what the customer wants, how this could be achieved, where there are strong relationships or conflicts between elements, a comparison of how selected competitive products achieve the goals and what the degree of technical difficulty is likely to be in achieving them.

The Japanese use these kinds of charts throughout their development exercises. Their culture seems to promote the use of small symbols to represent attributes and a desire to be able to display the whole picture of a particular thought pattern on to one piece of paper. QFD is a fine example of this and I believe has been used, not necessarily in its current form, in Japan for years without being called QFD until it was marketed to the West! Nevertheless, it is an excellent way of keeping the emphasis on the customer's requirements throughout the development cycle and focusing attention on the areas that bring most support to these requirements. At any point during the process a customer and supplier can be defined. The customer is the person receiving an input which is ideally exactly the output of the supplier. We tend to think of the customer as the person who buys the final product, which of course is correct. However, there are many customers and suppliers along the way. In broad terms, the marketeer is the customer

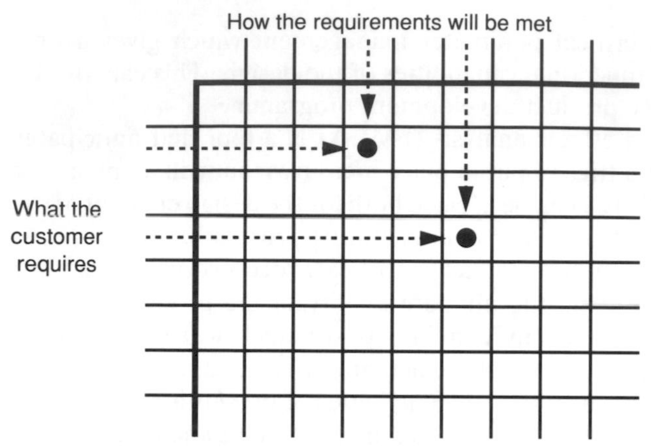

Figure 3.2 'What' versus 'how'.

of the designer. The designer is the customer of manufacturing. When drawings are being supplied to manufacturing, the designer is the supplier and manufacturing is the customer. Therefore the customer/supplier chain links all the way through the process from inception to launch and beyond.

QFD helps to define the relationships in the chain. At any stage it describes the matrix of *what* the customer wants versus *how* the supplier will supply it (Fig. 3.2). At the same time it identifies, using symbols, where there are strong relationships, medium relationships and weak relationships (Fig. 3.3). The core of QFD is to compare customer requirements with supplier specifications in a simple matrix.

The first customer requirements are those of the actual end customer, the person who will buy and use the product. The first supplier specifications

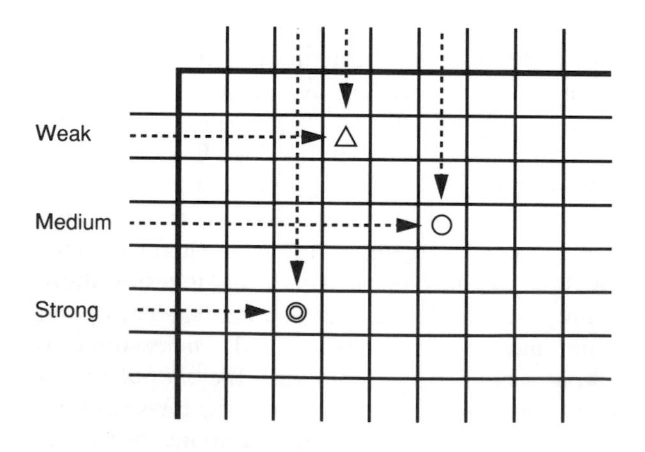

Figure 3.3 Matrix relationships.

are broad details of how these customer requirements will be met. For example, in the case of a car, some customer requirements may be:

List 1
- Low capital cost
- Good handling in wet or icy weather
- Long servicing interval
- Very low internal noise level

The response to these requirements in terms of suppliers' specifications may be:

List 2
- Standard engine and suspension design
- Front wheel drive
- New technology oil filter material and oil cooler
- Vibration-free engine mounting
- Extra sound insulation around passenger compartment

Some of these specifications may support more than one of the customer requirements. Some may satisfy one customer requirement while creating conflict with another. We will see how this is depicted later. The relation-

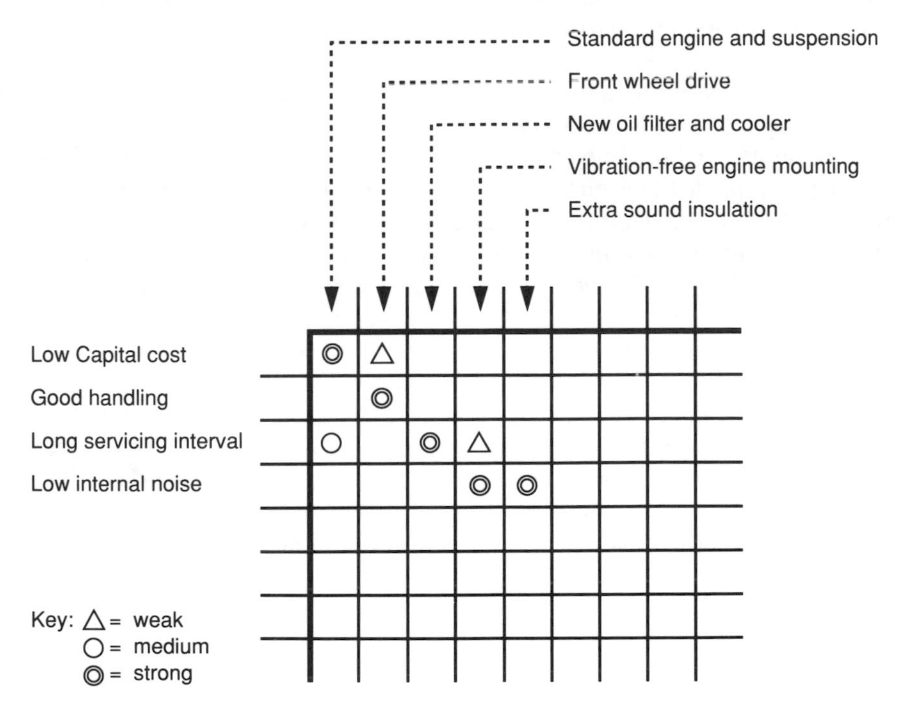

Figure 3.4 Example of 'what' versus 'how'.

ships between lists 1 and 2 are shown in Fig. 3.4, together with whether these are strong, medium or weak.

Now another similar matrix can be drawn, this time using the list of supplier specifications (List 2) as customer requirements. The supplier to meet these requirements might be the systems engineer and might respond with a list of systems specifications as List 3:

List 3

1200 cc and 1600 cc engines and transmission systems
Model 20 suspension system
Transmission to front wheels by double CVD system
Oil filter and cooler to new specification No. 2736
High hysteresis rubber engine mounting
High mass sound insulation material No. AS27365

Once the list of systems specifications has been identified as List 3, then these can be regarded as customer requirements for the subsystem engineers to supply, which would mean detailed specifications and parameters being identified as their supplier specifications (Fig. 3.5). This could be List 4.

With List 4 as the subsystem requirements, a list of supplier specifications to be achieved by manufacturing can be developed. This would be the specific process controls and quality standards that would need to be achieved to meet the subsystem requirements. This could be represented as List 5.

The final matrix would compare the requirements as defined by List 5 with the specifications of the finished parts to achieve the end customer requirements (List 6). In each case relationships can be identified and the distinction is made between those that are strong, those that are medium and those that are weak. Thus by linking each of these core QFD matrices into a system the process maintains linkage between end customer requirements and each of the steps in the process towards delivering the finished goods (Fig. 3.6).

Let us now look at some of the details of the core matrices used in QFD. Laying out customer requirements and supplier specifications in a matrix enables each of the customer requirements to be compared with each of the supplier specifications. In the body of the matrix symbols are entered representing the strength of the relationship between the two lists. The symbols usually represent relationships in terms of strong, medium and weak relationships. This creates a picture that might look like that in Fig. 3.7.

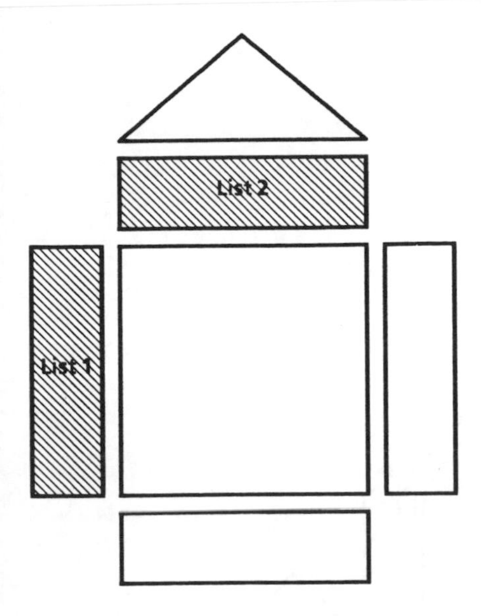

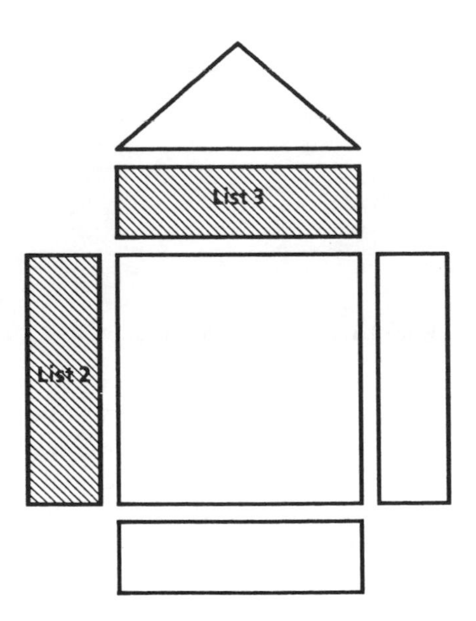

Figure 3.5 The linkage between matrices.

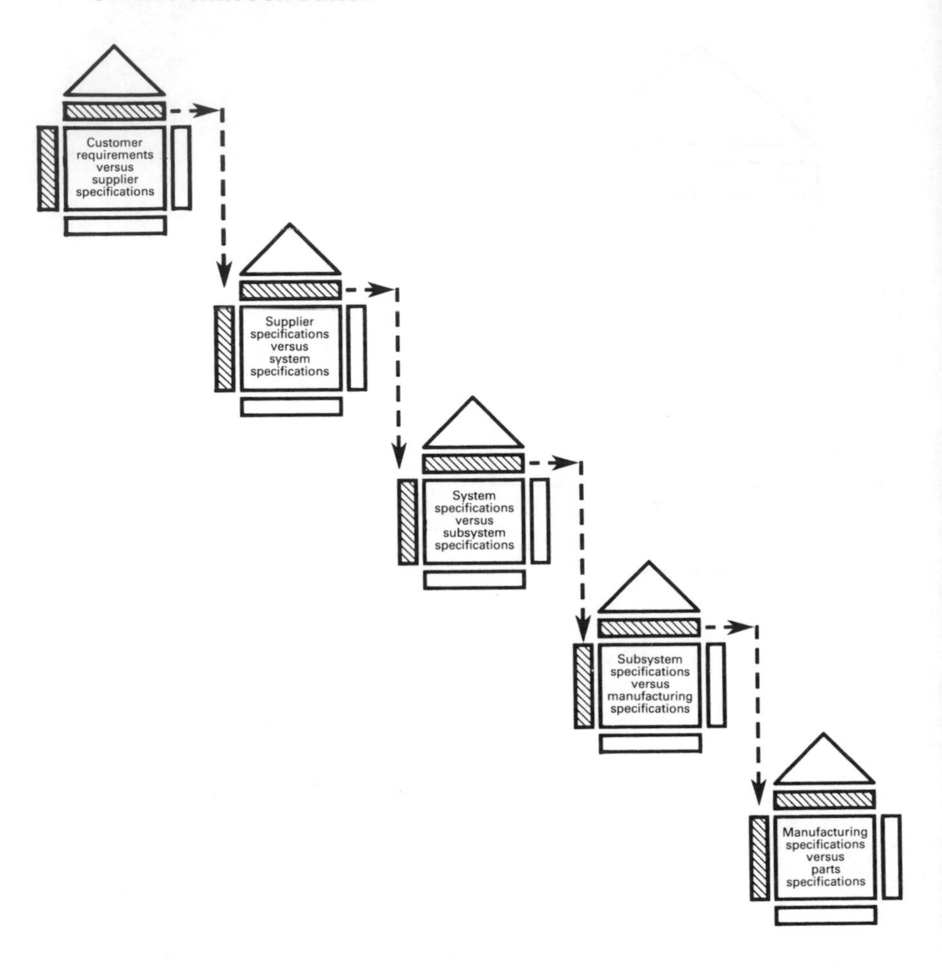

Figure 3.6 The series of matrices.

Another area that is very important to the user of the QFD matrix is where some of the supplier specifications are interrelated with each other, in terms of a positive or negative relationship. This can be depicted by filling in the 'roof' shape to the matrix (Fig. 3.8). Here the relationship must be depicted as:

Strong positive relationship denoted by ◎
Positive relationship denoted by ○
Strong negative relationship denoted by ✕ (bold)
Negative relationship denoted by ✕

A positive relationship between two factors means that they help each other in their interactive association, whereas a negative relationship means that they oppose or resist each other.

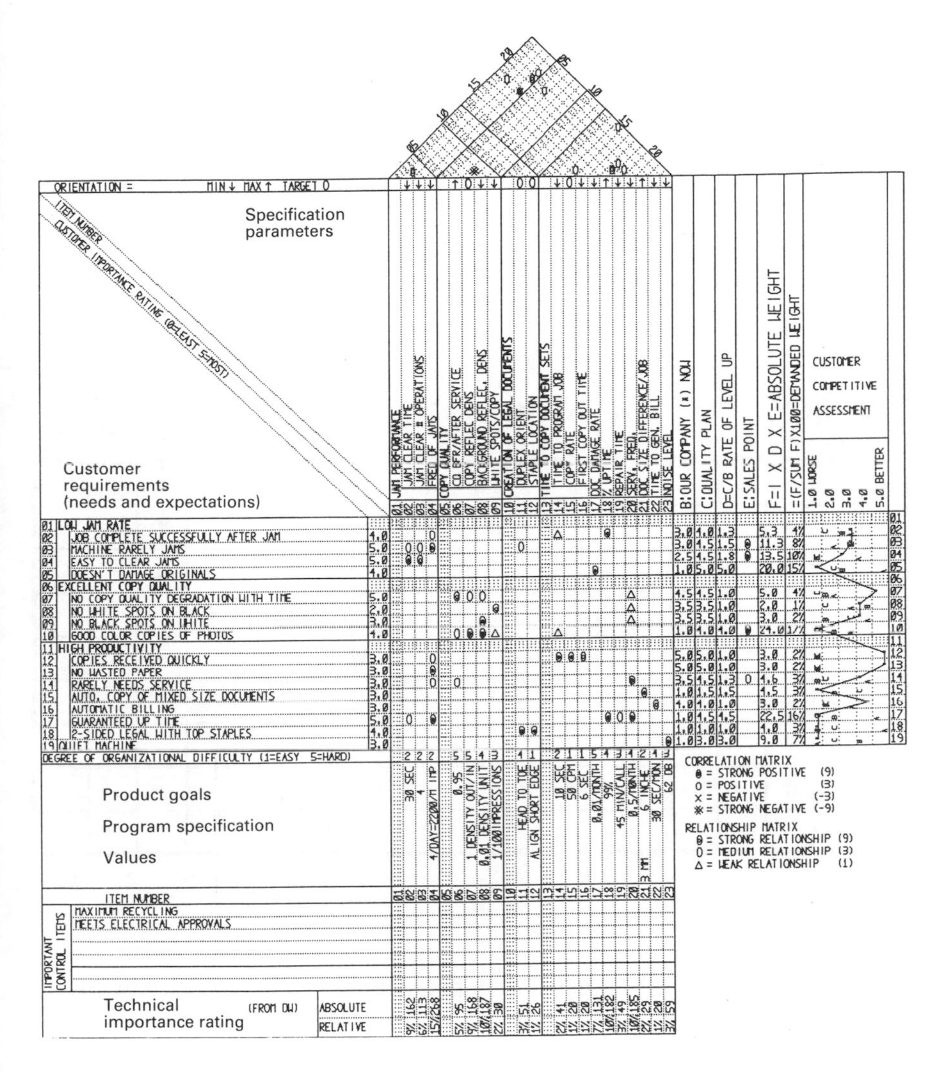

Figure 3.7 A typical QFD matrix.

The right-hand side of the matrix is where the importance rating and the competitive analysis are listed. From this information the prioritization of each of the items that has the greatest potential for success will be developed and used in the planning (see Fig. 3.9 on page 81). The bottom part of the matrix (Fig. 3.9) shows the prioritized supplier specifications and how difficult it will be to achieve these successfully.

Each of the matrices in the system can be built up in the same way as this. Once this has been done the users will have a readily accessible set of information showing which items are the most important to address, how

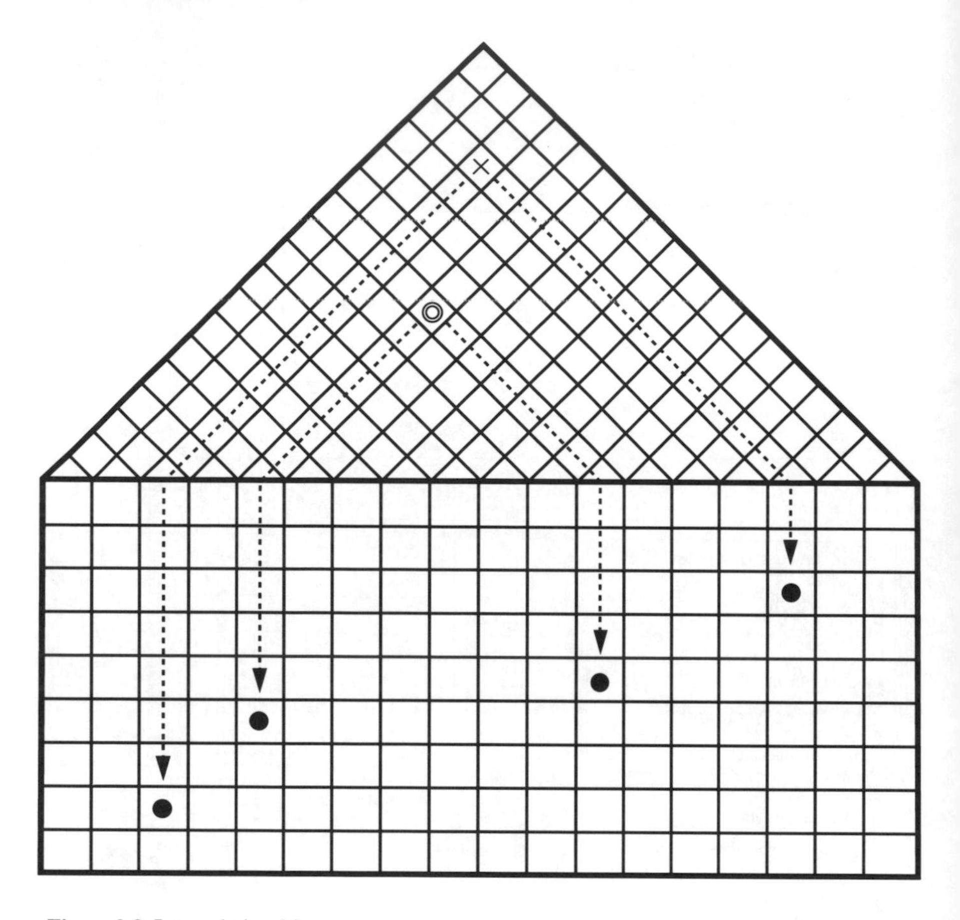

Figure 3.8 Interrelationships.

this compares with the competition, what technical difficulties are involved and how well the original requirements of the customer are likely to be met.

3.3 FAST

The study of function and its dependencies has in the past been neglected. It is essential to have a full understanding of function for each element of the design and, more importantly, to understand the interrelationships and dependencies of all the functions that form the complete system. There is a useful technique for studying design with regard to its functional relationships. The technique is called the *functional analysis system technique*, or FAST, and enables the engineer or designer to gain a grasp of the total design picture at any stage in the development process. This, as will be seen

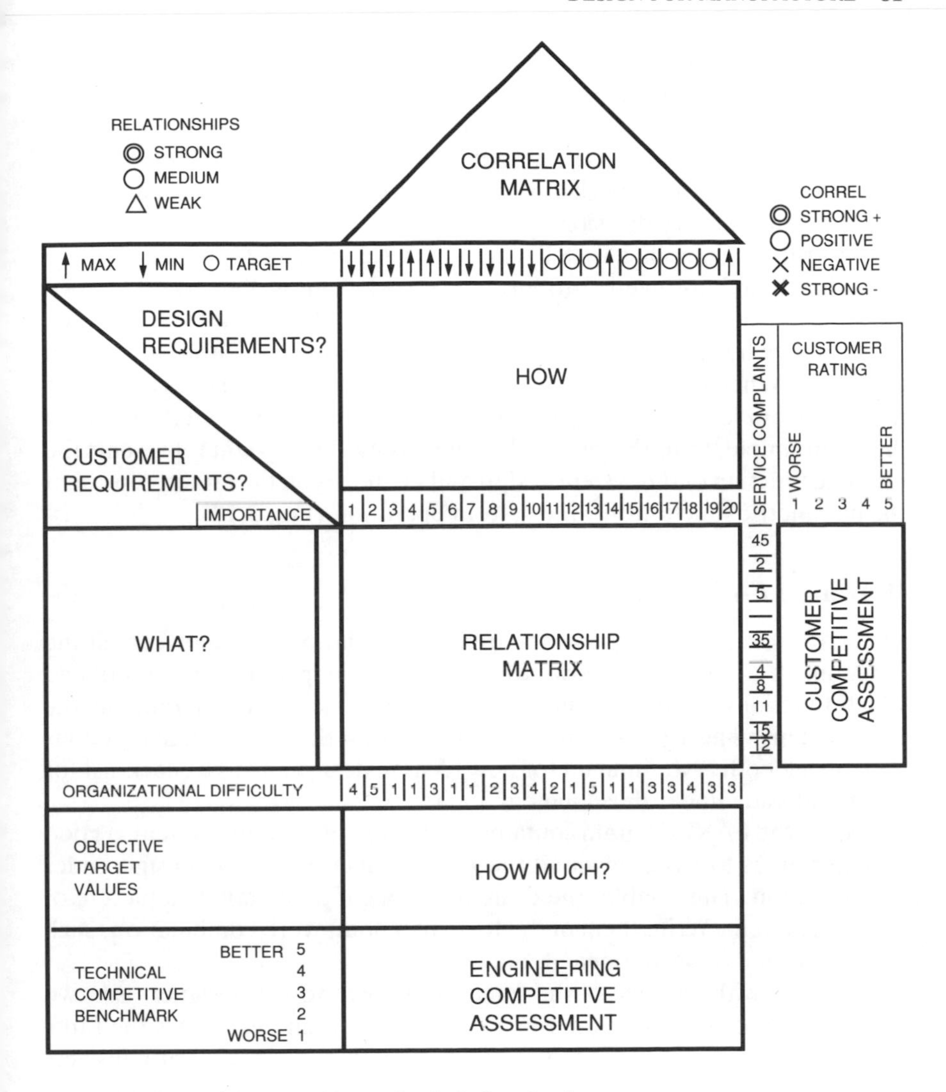

Figure 3.9 QFD with competitive and technical evaluation.

later, enables a thorough understanding of cost evaluation and reliability allocation throughout the various subsystems.

The process requires that each design function be identified and described by a noun–verb phrase. If one is in the concept stage of the design process, this immediately helps to sharpen the thinking on what the system design is intended to achieve and opens the door at an early stage to optimization of the very layout of the system, tending to drive the design towards its most simple form. In listing the noun–verb phrases for each function, one inevitably comes up against the question of how broadly should a function be regarded. For instance, the use of an electric motor to provide motive power can be assumed to be a single function, that is provide torque. In fact, of course, a motor contains many functions within it and is essentially a subsystem in its own right. This is an example of where one can regard a subsystem as a 'black box', and when one does or does not do so this has to be an arbitrary decision.

In order to compile a FAST diagram, it is necessary to link each of the individual functions together to form a network. This is done by asking two questions relative to each individual function: 'Why?' and 'How?' The answers to 'why' are placed to the right and the answers to 'how' are placed to the left. When the final network is completed, we can progress across the network from left to right by asking 'why' and from right to left by asking 'how'. In simple terms this means that the box on the far right represents the ultimate function of the system design and the boxes on the far left represent the parametric functions for the system operation (Fig. 3.10).

Drawing a FAST Diagram

The FAST diagram represents the functional relationships between all the functions of a system. It shows which ones are dependent on others and which functions occur at similar times or in parallel. The diagram has the benefit of bringing a greater understanding of how the total system functions and can act both as a stimulant for new ideas in design and as a check-list for designs already under development.

Since the FAST diagram contains all the functions of the system, a good way to start is to write down all the known functions of the design under consideration. This enables the designer to begin to assimilate a picture of the total system. Write them in the form of a noun–verb combination, such as shaft turned or valve opened, etc.

Starting with any one of the functions known to the designer, ask two questions. Ask 'why' does the function exist and then 'how' does the function occur. Answer each of these questions with more noun–verb answers representing other functions. It is a good idea to use sticky backed notes to write the functions on so that at a later stage the elements can be moved around into a more logical order as the diagram develops. Place the answers to all the 'whys' to the right of the relevant function and all of the 'hows' to the left. Generally there will be more answers to the question 'how' as this breaks down the make-up of the function in question. For instance, to turn shaft might require apply torque, provide bearing, etc. Link the functions together with lines so that all the necessary 'hows' link into the function that they enable. Once more functions have been added to the diagram; repeat the 'how' and 'why' question format for each one and allow the diagram to expand (Fig. 3.11).

Sometimes it will be noticed that the same verb–noun function is

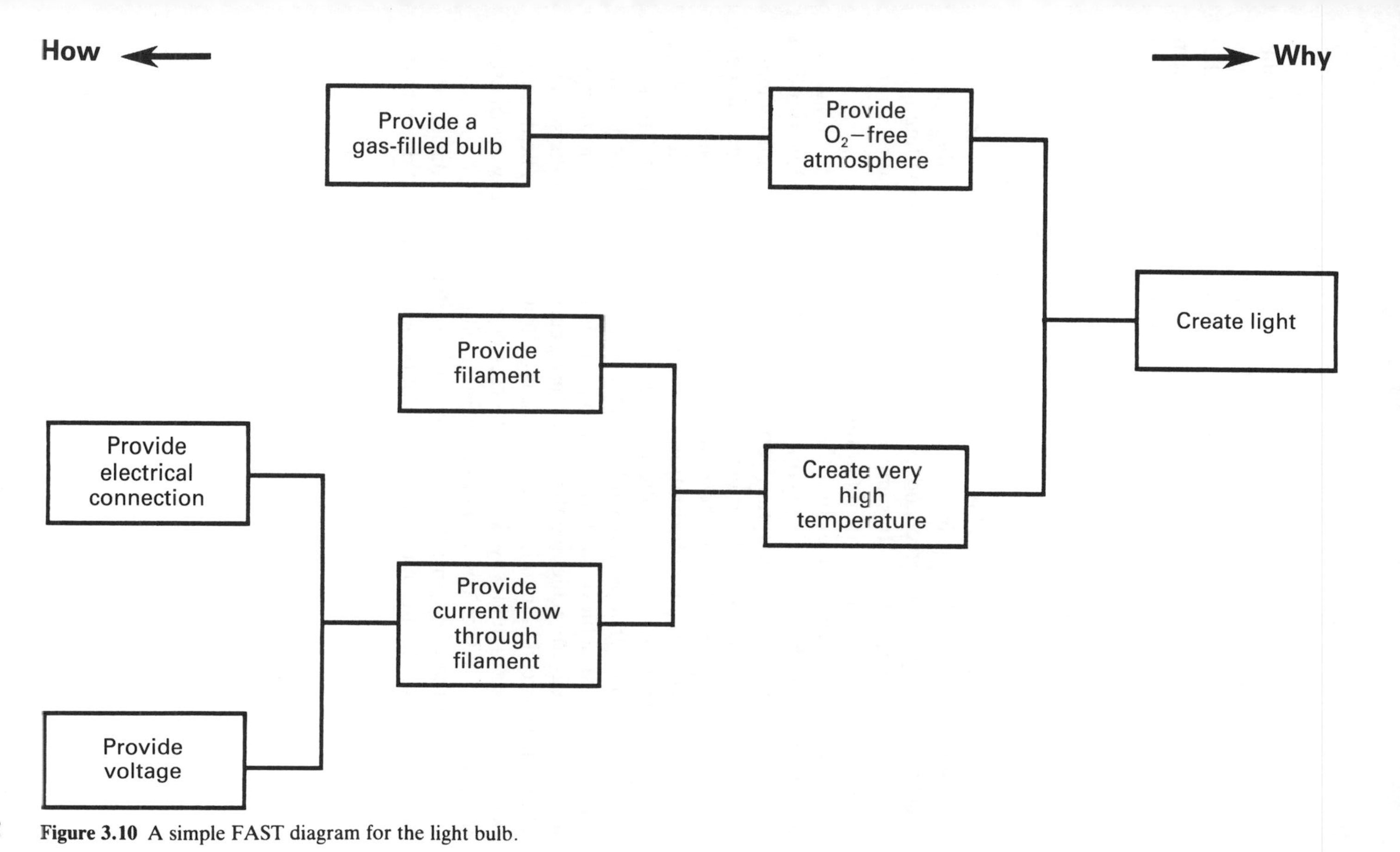

Figure 3.10 A simple FAST diagram for the light bulb.

repeated. If both of these represent the same function in the design then only write it once, but link it to both parts of the diagram that this enables. If they are just the same noun–verb combination but represent functions that happen separately or in different parts of the design, then keep them separate and write them down twice or as many times as they occur independently. This will be beneficial later as it will draw attention to functions that are similar but are addressed separately.

Keep working the diagram asking the 'how' and 'why' questions until you are satisfied that the diagram is a true representation of the system function. One of the most usual errors is to answer the questions with too all-encompassing an answer. For example, 'How is the valve opened?' By applying force, providing spring and providing guide. In fact 'valve opened' is achieved by 'applying force' and 'providing guide' only, 'providing spring' being 'how' we apply force. It is important to break down the system into its smallest elements to make the diagram most effective.

It will soon become clear that the functions at the left of the diagram, usually prefixed 'Provide . . .', actually represent the parts, while those on the extreme right of the diagram represent the customer's requirements. The completed diagram can be used immediately to gain a broad understanding of the elements of the system. Look for areas that seem overcomplex for the function contribution that they make. Look for areas where functions are repeated and ask whether these could be combined. Break sets of functions into groups and study whether these appear to be the most efficient way to achieve performance or whether this gives the designer alternative ideas for achieving the same. Just a cursory look at the structure of the diagram can help to initiate new ideas of how the job can be done in a better or more efficient way. Later I will describe more formal ways of using the diagram to reduce costs and improve reliability.

In using the diagram it will be necessary to define boundaries of the work. For example, in some tasks it will not be necessary to get down to the

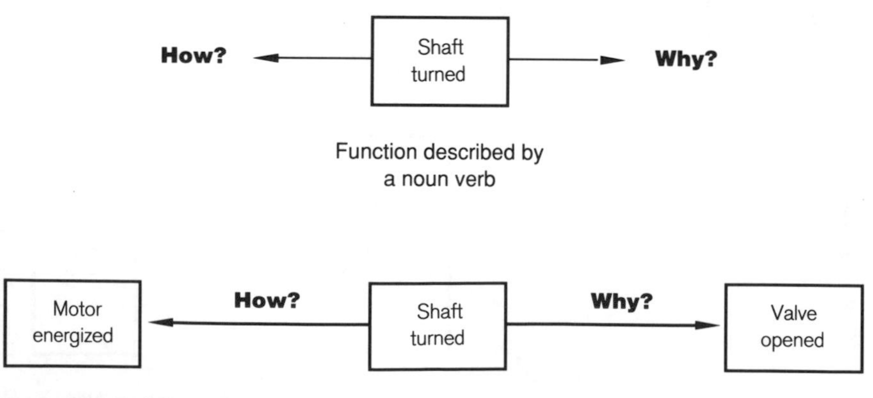

Figure 3.11 Building a FAST diagram.

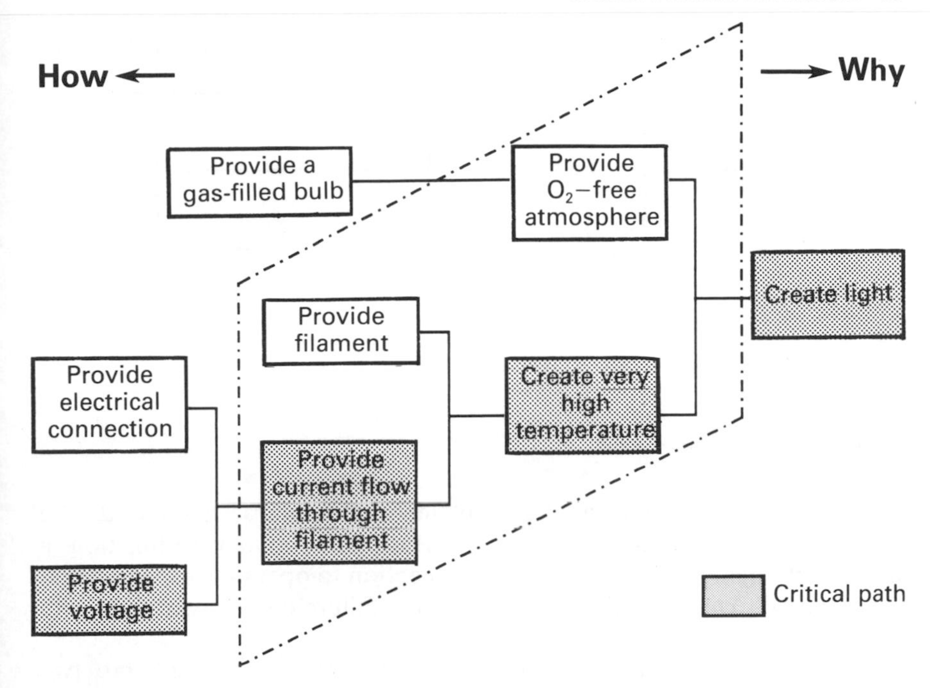

Figure 3.12 The FAST diagram for the light bulb showing critical path and boundaries.

parts level, so a boundary can be drawn to the right of this. Similarly, if ultimate customer's requirements do not come into the exercise being undertaken a line to the left of this can define the boundary.

It is often useful to draw the critical path through the diagram from left to right. This line represents the central functions of the system at all levels (Fig. 3.12).

3.4 THE FUNCTION DIAGRAM

Having prepared a FAST diagram, the designer has a clear view of the overall system design and can begin to look at this structure to see, firstly, if it seems to be the most efficient arrangement that can be configured and, secondly, to highlight the most significant functions of the system. It will usually be the case that there are just a few key functional areas to the system. (Indeed, if there are a large number this tells the designer that the system is too complex to be dealt with in its present form without further breakdown, and enables the process of breaking down the system into smaller components to be studied.) For each of these key functions a function diagram can be configured (Fig. 3.13).

The function diagram is the heart of the process called critical parameter management (see Sec. 3.5). It provides the foundation upon which the

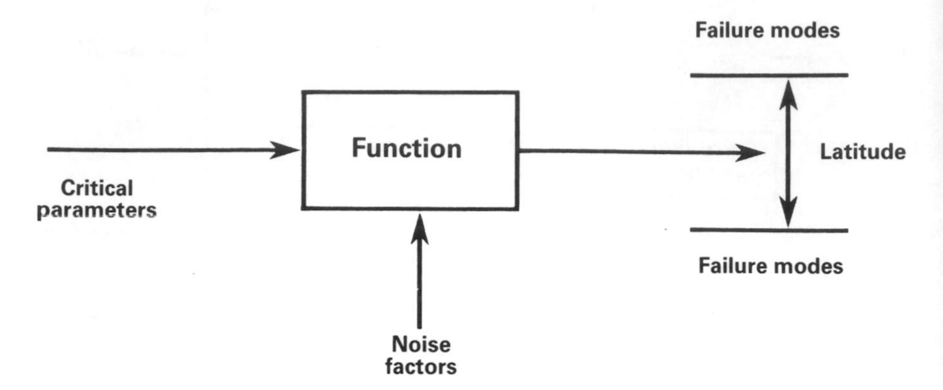

Figure 3.13 The function diagram.

understanding of latitude and the enablement of optimization is based. The function diagram comprises a box representing the noun–verb function, a list of input parameters that enable the function to operate, responses that may be used to measure the function and the failure modes associated with the function.

The parameters that control the function can be divided into two groups: those that are within the range of control of the designer, and those that are outside the designer's control. Those that are within the control of the designer are the only means by which the designer can cause the function to operate in the way that is required. They will generally be described ultimately as all the dimensions on the drawings that go to manufacturing to produce the design. But the dimensions on the drawings are really the working parameters for manufacturing, not the fundamental parameters the designer requires to be held in order for the system to work. For example, if a system needs a force to be applied to open a valve, it is not normally this force that is specified ultimately on the drawing. More likely it will be a series of dimensions describing the parts and their interface with the surrounding parts to enable the correct force to be delivered. In preparing a function diagram at this stage of the design it is the fundamental parameters that are required to be listed rather than the final embodiment of them in terms of dimensions.

Thus the function diagram is built up by firstly defining the verb–noun function (Fig. 3.14). From this the response expected from this function is defined and should be a factor that can be measured. The response is considered from the customer acceptance point of view and upper and lower failure limits are defined. Finally the parameters that control this function are added: those in the designer's control (critical parameters) and those outside the designer's control (noise factors).

In preparing the list of functional parameters the designer should think in terms of fundamentally enabling parameters and not the dimensions that are required to manufacture the design. The list may be fairly long at this

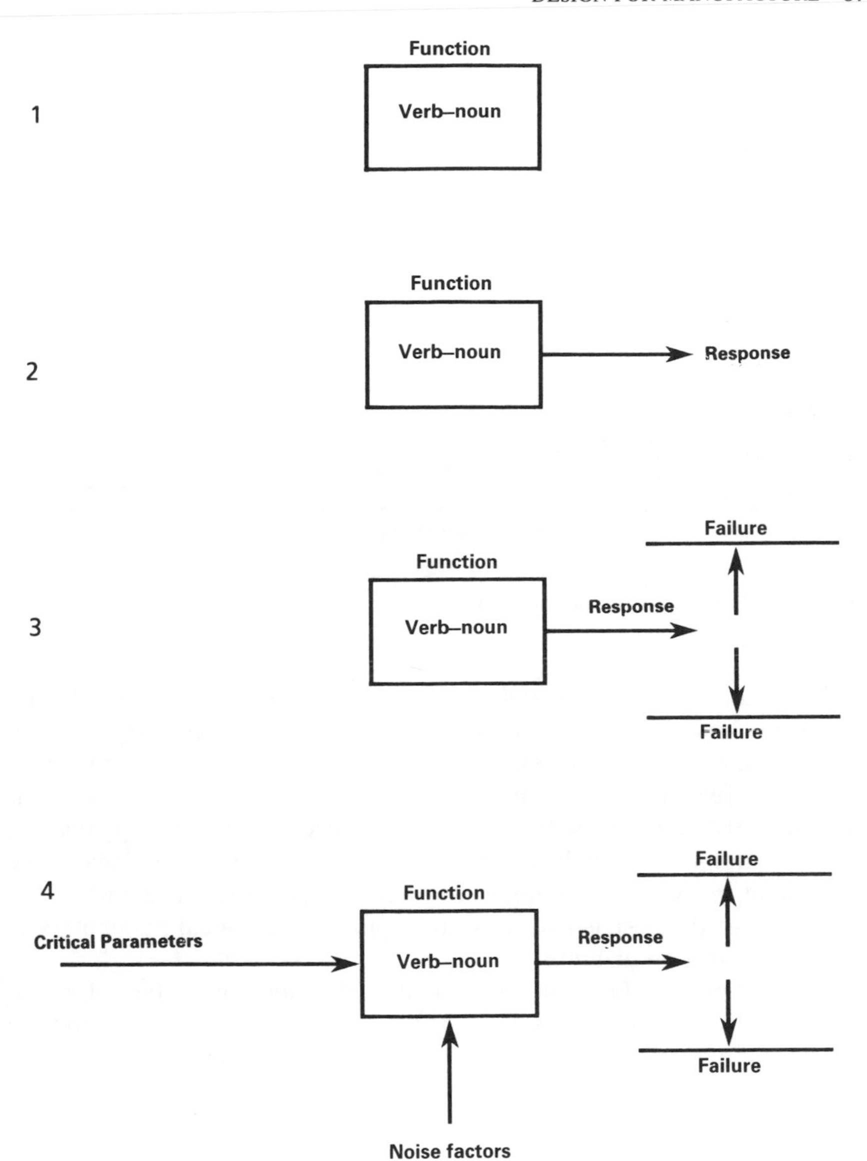

Figure 3.14 Building the function diagram.

stage and so the next step is to try to identify which of the parameters are 'critical' to the function. Here the word critical needs some definition. If one were to plot all the parameters to show the order of sensitivity to the response of the function the curve would look something like Fig. 3.15; that is parameter number 1 on the left of the chart is most critical to the function

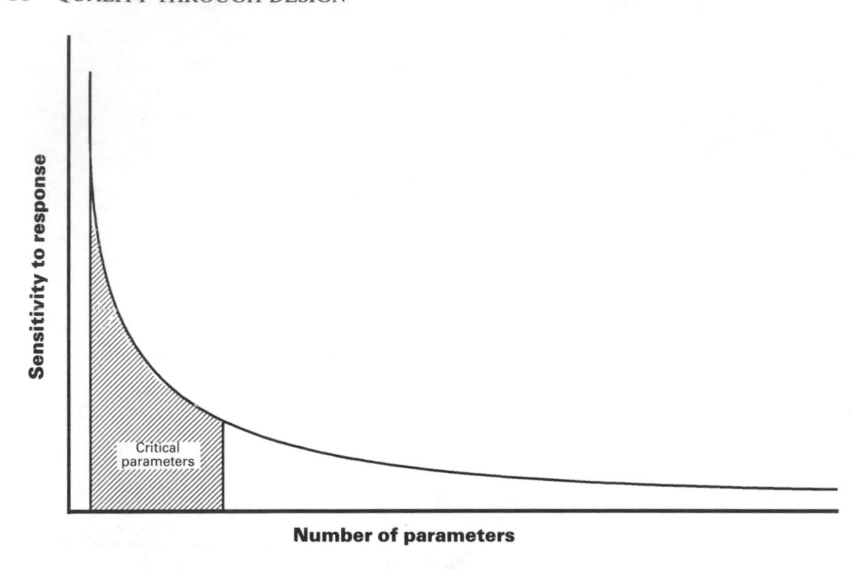

Figure 3.15 Definition of critical parameters.

whereas parameters on the right of the chart are much less critical. The right-hand boundary for the critical parameters is an arbitrary decision drawn from experience. In general, the group of critical parameters represents about five or ten per cent of the total but this will vary with different technologies and products. Precision in the selection of parameters that are critical is not important because if a bad selection has been made the subsequent process will highlight the shortfall. It is therefore part of the judgement of the designer to select the appropriate critical parameters as part of the function diagram.

The second set of parameters that affect the function are those that are outside the control of the designer. These are called noise factors, and are usually environmental factors such as temperature and pressure. It should be noted, however, that when one is considering a system the output (or response) from one subsystem may be the noise factor for the next. For example, the range of torque output and its variation from a motor will be a noise factor for the gearbox that it drives. In other words, the designer of the gearbox must accept the variation of the motor torque as a given when considering the gearbox design and either accept the effect that this has on the gearbox performance or design in features to compensate.

In all the input parameters, whether control or noise factors, it should be appreciated that each one of them has a nominal value and a range. During the process of critical parameter management, both of these factors must be considered.

Each function exists to provide an output. It is this output that the designer requires to combine with other outputs of the system and provide the final operational function of the product. The output for the single function under consideration will appear as some form of response. This will be a force, a displacement, a torque or similar output. In order for the designer to evaluate the effectiveness of the function this response will need to be measurable, and having measured it, the designer can decide whether it fits within the acceptable range for this response. The requirements of the customer will ultimately set the limits of this response. When the response exceeds either the upper or the lower limit set, the function is said by definition to have failed. Thus, by looking at responses and their upper and lower limits, one can also define the failure modes of the function. Bear in mind that a response will probably have more than one failure mode at each of the upper and lower limits, and indeed there may be more than one response to monitor in order to evaluate the function.

In many ways the definition of 'failure' in this sense is a definition of the customer's requirements not being achieved. For example, take the response to a function of acoustic noise. The response will be said to have failed once the noise level exceeds that which can be tolerated by the customer, and this will be dependent on the nature of the application of the product. A refrigerator working in a domestic environment will need to have a low limit of acoustic noise tolerance, whereas a lawn mower could be allowed to reach a fairly high noise level without becoming intolerable. Thus the definition of failure and the failure mode depend on where we set the limit of tolerance, which, in turn, depends on the application and the requirements of the customer. This concept of failure contrasts with the idea of a breakdown or catastrophic failure such as the breaking of a part.

The failure modes that are associated with the function and its response are in a sense layered. As an example let us consider the function of the inflation of a car tyre (tyre inflated) (Fig. 3.16). This is measured by the response 'air pressure'. The latitude of air pressure in this particular tyre may range from 20 to 25 lb/in^2. What are the failure modes at both ends of this latitude scale? As pressure increases the tyre gets harder and grip is impaired. As the pressure increases further the ride becomes uncomfortable, then wear increases, then damage to the structure takes place until finally the tyre bursts. These failure modes form layers of failures which gradually get worse until they finally become catastrophic. As pressure decreases, there is a point where the driver notices some degradation in the handling of the vehicle, the first failure mode. After this further reduction in tyre pressure may result in a definite wobble in the steering, then inability to control the vehicle and finally the catastrophic failure of irreparable damage to the tyre.

This is an example of how the response from a function can be assessed in terms of its ability to satisfy the customer's requirements in terms of performance. Knowledge of this and particularly its relationship with failure

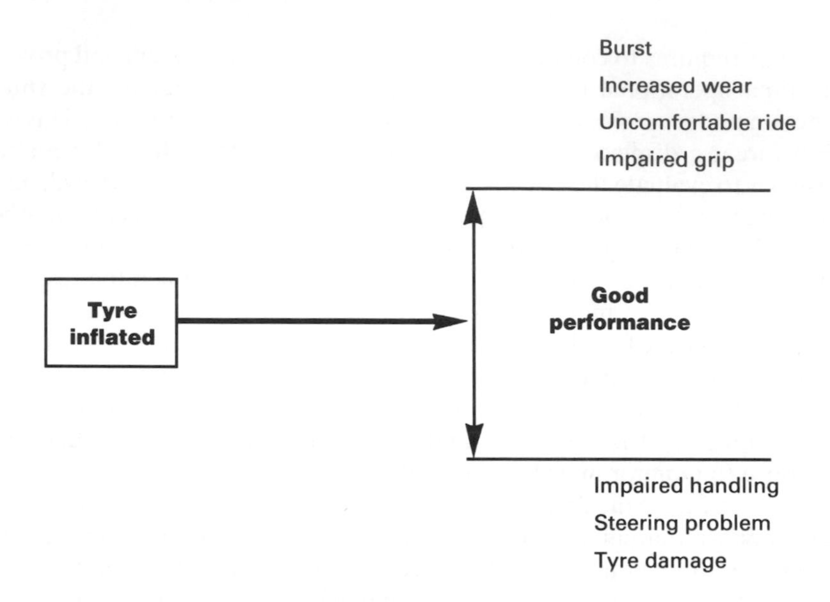

Figure 3.16 Partial function diagram for a tyre.

modes helps the designer to gain a clearer understanding of how to achieve a robust design and also how to test it.

If the designer understands the sensitivity of each of the critical parameters to the response from the function, then expanding latitude becomes a process of minimizing the variation of those critical parameters that have the greatest effect on the functional response (Fig. 3.17). If this means closer control of the particular critical parameter then the designer can choose whether this is the expedient thing to do. If it turns out that the critical parameter cannot extend the latitude of the response sufficiently then it may tell the designer that the chosen technology is incapable of delivering the required level of robustness for the particular application.

It may be that the sensitivity analysis shows that some of the critical parameters do not have a significant effect on the latitude of the response. If this is the case then the designer may choose to try to reduce the cost of the part by allowing the critical parameter to vary more widely.

All these possibilities become the choice of the designer once an understanding of the function diagram is obtained.

During the preparation for testing, a study of the function diagram, and in particular the latitude of the response, can be a great help in designing tests that will truly identify the reliability of the product. If the actual level of the response is measured so that the test results show the variation of the operating point for this function, the engineer can decide just how close to failure the product is running. On the other hand, if only failures are

Critical parameter tolerances	***Latitude***
Increase to reduce cost	Wide
Review failure modes Reduce tolerances New Manufacturing Process More cost Consider capability of chosen technology	Narrow

Figure 3.17 The relationship between latitude and critical parameter tolerance.

recorded during a test, it is easy to see that a situation could arise where the product is operating very close to failure but appears to be running all right. Any slight variation in a parameter could cause it to deviate sufficiently to cause failure. The test would have to run for an enormous length of time to provide the same information as a well-constructed test measuring response level.

It is important to understand the relationship between engineering and manufacturing in this context. When the development engineer starts out to develop the design for the system certain parameters are controlled to achieve the required performance. The better these parameters can be controlled the more consistent and robust will be the output performance of the product. However, this is not a good track to follow, for although the performance can be improved by tightening tolerances, the product will be easier to manufacture and more resistant to changes if the tolerances set by the development engineer can be as wide as possible. Therefore the engineer strives to make tolerances as broad as possible to achieve this; that is every parameter needs to be 'tolerant' or not particular about a very accurate value.

The manufacturing engineer, on the other hand, would like the processes that are used to make the parts to deliver very accurate parameters, because it is known that the less variance that is evident in the parameters of the parts the more likely they are to meet the tolerance that the design engineer has set. Therefore the situation occurs as time passes that the design engineer starts with very close tolerances and tries to increase them as much as possible, whereas the manufacturing engineer starts with very wide variances in the manufacturing processes and tries to reduce them as much as possible.

Once the manufacturing variances sit within the design tolerances the product performance is robust. (Refer to Fig. 2.14.)

Definitions

Control parameters These are the parameters that the designer uses to effect the function required of the system. They are the parameters that are within the control of the designer and that affect the function. They are also the parameters that are varied during the experimental work.

Critical parameters These are the control parameters that are critical (or paramount) in their effect on the function. For any subsystem they are expected to be about 10 per cent of the overall set of parameters. The definition of 'critical' is arbitrary, but it is based on the premise that if a critical parameter goes out of the range prescribed to it in the design, then the function is likely to experience a failure.

Noise factors These are factors (parameters) directly affecting the function which are outside the control of the designer.

Responses These are the outputs from the function which are measurable. They may be noise factors for the next function within the system.

Failure modes A failure mode occurs when the response from a function exceeds (either positively or negatively) the limit expected from the design. It may not necessarily be a hard failure (that is one that causes a breakdown of the system), but may be manifested as a failure to meet the requirements of the customer. In this way failure modes can be 'layered', or brought in progressively as the response drifts further from the prescribed range.

Latitude This is the full range of response as described by the failure modes. In other words, it is the satisfactory working range of the function.

The Function Diagram and Its Importance

The function diagram can be considered to be the keystone of the design process. It ties together the relationships between the parameters that control the function, the function itself and the outputs from the function and their associated latitude and failure modes. If the designer can get a full understanding of these relationships, then the ultimate design quality and the rate of progress of the design itself will both be enhanced.

Let us consider some of the valuable information that can be obtained from a fully developed function diagram, bearing in mind that the design has not at this stage been worked on in the traditional way of preparing drawings. Firstly, the designer can look for narrow latitude ranges in the responses from the function. If a range is too narrow (that is the response has very narrow latitude) the designer immediately knows that this is a situation likely to incur failures. Is there anything that can be done to provide wider latitude? If the critical parameter (or parameters) that affect this particular response is known, steps can be taken to modify this parameter to provide a wider latitude. This of course requires a clear understanding of the relationships between critical parameters and responses. Taguchi methodology provides this understanding.

If a response has wide latitude the designer can consider changing the critical parameter that affects this response, and this may provide an opportunity to reduce costs. Looking at the list of critical parameters and their proposed nominals and tolerances, the designer may decide that a certain parameter requires too tight a control in manufacturing to achieve the desired functional response. This will highlight the fact that there is a problem in the manufacturing of this element (it is either too costly or requires a new manufacturing technology to achieve). Attention is therefore immediately drawn to the question of how can this be resolved, either within manufacturing or by changing the design approach to eliminate this particular problem.

In looking at the responses identified and their associated failure modes, one can determine the most effective method of testing the hardware once it has been designed and built. For example, it is important to test for the actual responses, not some effect that they have. It is also important to establish by testing exactly where in the latitude band one is operating, not just when failures occur. A good example of this is in the testing of paper handling systems for copying machines. Copying machines rely very strongly on a reliable system for separating, feeding and transporting sheets of paper through the system. Testing these systems has often been done by testing for the number of times a sheet of paper jams or fails to progress through the machine, and the failure rate would be expressed as so many jams per million sheets. The ineffectiveness of this can be seen if one imagines a system operating just on the threshold of failure. A test measuring jams would provide a good result under these conditions whereas in fact the system would be operating in a very vulnerable condition. If instead the 'tendency of the sheets to stop' were measured by comparing the expected position of sheets versus the actual position of sheets, the result would be more representative.

Failures are by definition the condition when the function of the system does not meet the requirements of the customer. Here we have another area in which we can exercise some control. For example, suppose we are designing a spin dryer and the response from the drive system is a certain nominal speed. If the speed is increased above this nominal value a failure mode might be to create excessive noise and if the speed is decreased the failure mode might be to give a reduced level of drying. During the development of customer requirements, the designer can negotiate the best compromise between customer and engineering to meet the most favourable settings of these limits. The point here is that the failure levels are often points that can be set by the designer and customer combined. They are not something that is completely outside our control.

Critical parameters have nominal values and ranges to these values. The nominal values are set by the designer to deliver the optimum performance of the function, and this is often the result of carcful and sometimes lengthy experimentation. The ranges about these nominals are something entirely different. They are the tolerances that the designer can allow while still maintaining the performance of the function within acceptable limits. As such they raise a number of important questions. The process of finding what is the acceptable range for each nominal critical parameter demands that the design team understands the relationship between each of the critical parameters and the functions that they enable. This requires knowledge of the function as part of a system and can be achieved either by analysis or by experimentation. Once these relationships are known in terms of their effect on the responses, an estimate of how far these nominals can be allowed to drift before causing an adverse effect on the response is required. Again this can be done effectively by employing Taguchi methods. Having defined the ranges, the designer can look at them in two respects. Firstly, how difficult will it be to hold these limits during the manufacturing process and during the life of the product? If the range is very narrow the designer must be concerned not only for whether the processes that will be employed in manufacturing the part or parts will be capable of doing so, but also whether during the life of the product, the range could be exceeded due to wear or some environmental factor. Conversely, if the range is very wide indeed the designer must decide whether this parameter is indeed critical or whether as a result cost should be reduced by employing a more coarse production method which will still achieve the required design tolerances but perhaps at less cost.

Here it is useful to define the use of the word 'tolerance'. Tolerance should only be used when referring to 'what can be tolerated in the design'. Limits defined in a design are what the design can tolerate and are therefore called tolerances. When talking about the limits that a manufacturing process can achieve one should talk about 'manufacturing variances'. There is no such thing as manufacturing tolerance.

Noise factors also have ranges. For example, a noise factor for a particular function may be temperature. The designer needs to identify the exact range of temperature variation to which the system is likely to be subjected. These data (nominals and ranges) must be used during the optimization experiments which go on when identifying the relationships between the critical parameters and the functions. Remember that these parameters, which may be critical, are outside the control of the designer.

Example of a FAST Diagram

As an example of a FAST diagram to show the functions of the system and their relationships, let us use the office stapler as the design under study. The only difference with this exercise is that in this case the design is already known whereas normally the design under consideration would be in an embryonic form and would need refinement and modification. Using an example of something that we are reasonably familiar with enables us to appreciate and understand the method more easily.

An office stapler (Fig. 3.18) consists of a base which supports a pivot to give freedom to the stapler assembly. The stapler assembly contains a slide to house the staples, a plunger to drive the staples and a hand pad to enable the user to apply the force to drive the staple into the sheets of paper. An anvil attached to the base and in line with the plunger allows the staple to form a clinch after the legs of the staple have passed through the paper and force is continued to be applied to the hand pad. The staples are supplied in 'sticks' which can slide on the slider and are biased towards the plunger by a spring.

A FAST diagram represents the complete functional operation of the system by linking the functions one to another such that each adjacent pair or group represents the true relationship between them. This functional

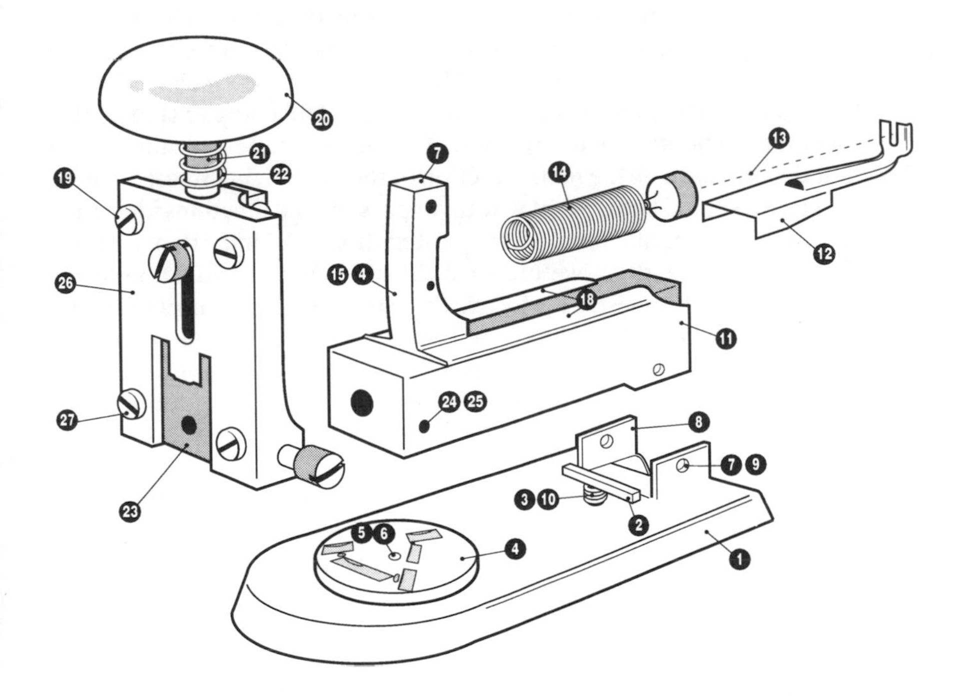

Figure 3.18 A desk-top stapler.

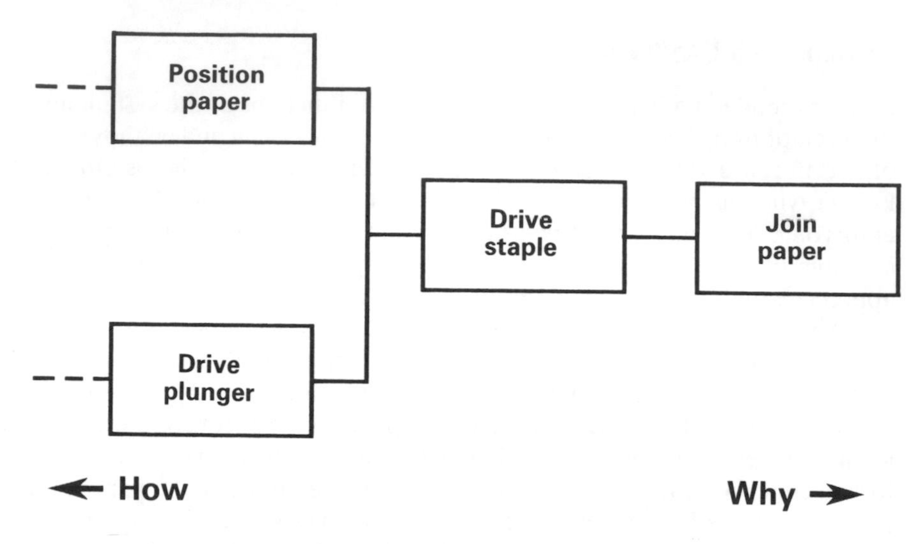

Figure 3.19 Starting a stapler FAST diagram.

linkage can be determined by asking, firstly, why the function exists, and, secondly, how that function is achieved. In order to draw a FAST diagram we must try to place the functions resulting from the question 'how' to the left of the original function and those resulting from the question 'why' to the right. In this way a diagram of functions that are linked together as a system is developed forming the complete FAST diagram for the system.

Let us start to develop a FAST diagram for the office stapler (Fig. 3.19). The diagram can be started at any point. Usually it is easiest to start with a function that seems to be the main function of the system, in this case, 'drive staple'. Here it should be said that wherever possible the functions identified should be in the form of a noun–verb combination. The importance of this will be seen more clearly as we progress. Starting with the function 'drive staple' we now ask the question 'why' we drive the staple. The answer clearly is to 'join paper', so this function is placed to the right of the first one and linked by a line. Next we ask the question 'how' we drive a staple. Here the answer is more complex. Firstly, we must have a staple in a position to be driven—'position staple'. Secondly, we must push the staple by some means—'drive plunger'.

Let us now go to the other function 'join paper' and again ask 'how'. The answer in this case is threefold. We must first 'position book' (for the staple to drive through), then 'drive staple' and finally 'clinch staple'. At this point we should also ask why we 'join paper'. Clearly the answer to this is that this is our customer requirement and we see that we need go no further to expand the FAST diagram to the right.

With this part of the diagram now completed we can continue to ask the question 'how' for the functions linking to the left. As the diagram is built up

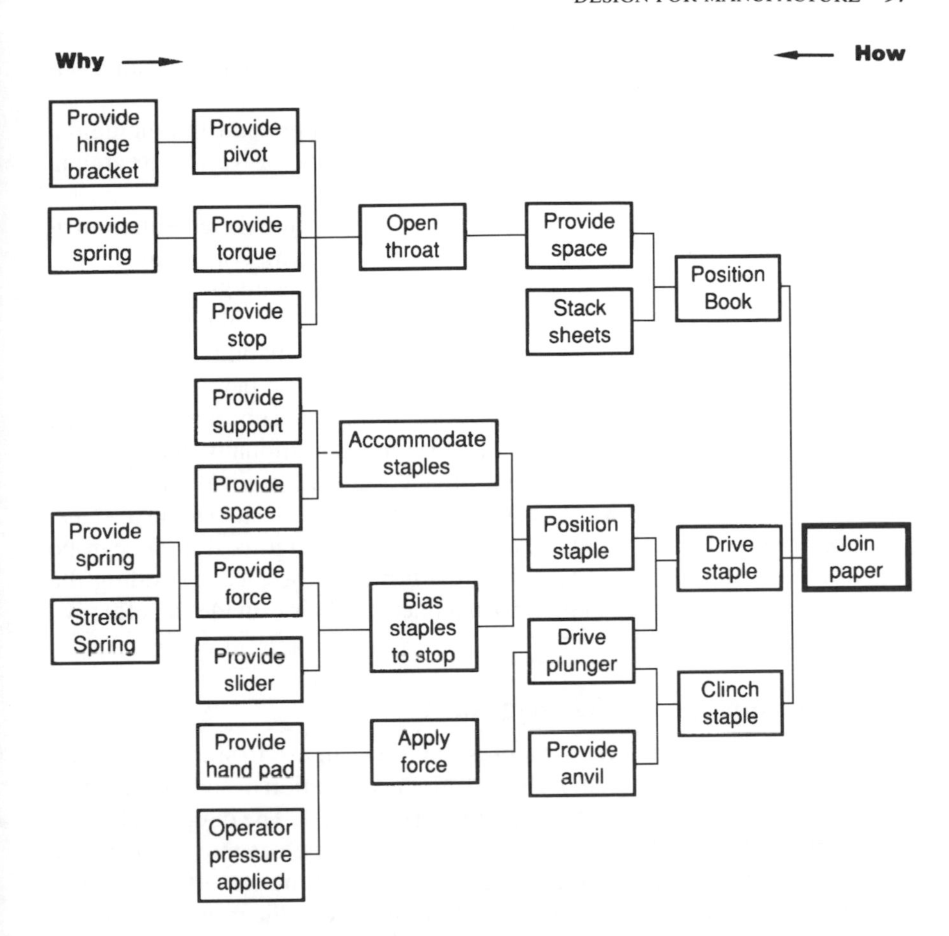

Figure 3.20 FAST diagram for a desk-top stapler.

to the left it also expands vertically as more and more functional dependencies are built on. These functions are also seen to be more detailed in their nature and as we extend the diagram further and further left it is clear that the actual parts start to become defined as a part of the functional description. We see nouns like spring, support, stop, etc., all of which are parts in the assembly. Thus it can be said that the left-hand extreme of the diagram depicts the individual parts of the system, while the right-hand extreme of the diagram represents the customer requirements (Fig. 3.20).

This diagram has been drawn with full knowledge of the design for reason of explanation. Normally the diagram is developed based on an idea for a functional system and can be invaluable to the designer as a tool to develop these ideas, refine them and work towards an optimum system relationship. However, it can also be used at virtually any time during the design process to clarify and optimize the relationships between functions and to perform value engineering.

Another valuable use of the diagram is to understand the elemental build-up of both reliability and cost. Both of these factors can be treated in the same way.

Using reliability as the example, each of the functional elements contributes in some way to the unreliability of the system. For example, the function 'position staple' will have a failure rate associated with it, and by adding each of the elemental parts of the failure rates a total failure rate for the system can be determined. With this picture of reliability the designer can easily focus on those areas that are significant and be aware of those areas that are negligible. Design changes can then be concentrated on the areas that are likely to bring the greatest return. Additionally, a design can be compared effectively with a competitive design and equivalent functions matched to see where the weak or strong points occur in each of the designs.

Cost can be treated in exactly the same way. High cost areas can be identified and the value of the individual function assessed by adding the costs of the parts that go to make up this function. This provides an effective approach to value engineering.

Second Example of a FAST Diagram

Probably one of the simplest examples that one can use to study the implementation of a FAST diagram is the functional operation of something most of us use every day, that of a ball-point pen (Fig. 3.21). The diagram shows a section through a simple ball-point pen. It is useful, before moving on to the implementation of a FAST diagram, to think of the functions of a ball-point pen and then later we can check to see that the FAST diagram has covered all the operational attributes. The following list summarizes the functions in no particular order of priority:

1. Meter ink on to the paper
2. Provide means of holding the pen for writing
3. Contain an adequate supply of ink
4. Prevent ink spillage

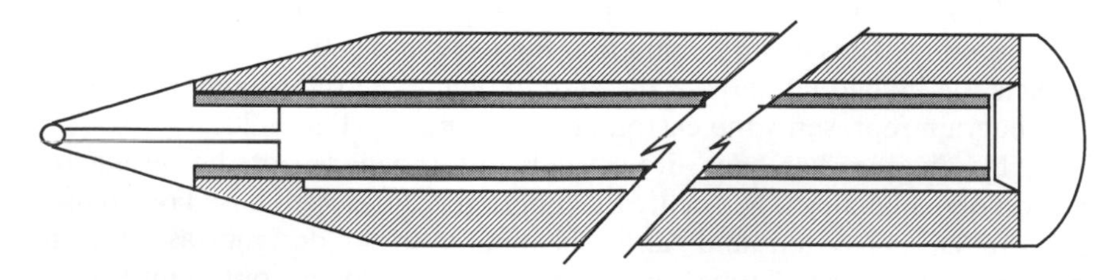

Figure 3.21 A ball-point pen.

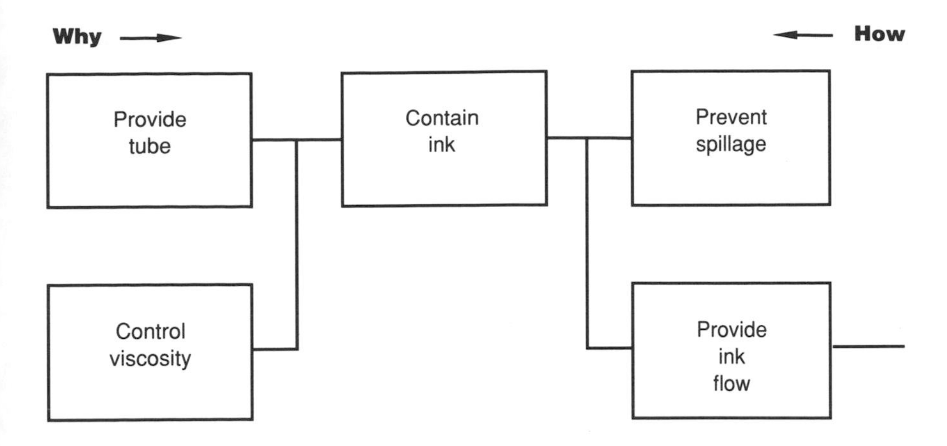

Figure 3.22 Developing the FAST diagram for a ball-point pen.

5. Allow the flow of ink
6. Protect the point from damage/inadvertent marking (optional)

We can construct a FAST diagram by starting with any obvious function and asking, firstly, 'why' this function is required and, secondly, 'how' this function is achieved.

Let us start with the function 'Contain an adequate supply of ink' (Fig. 3.22). For the purposes of constructing a FAST diagram we need a verb–noun phrase that describes this function. This would be 'contain ink' for this particular function. Why do we want to contain ink? The answer is to prevent spillage and to provide ink at the ball point. How do we contain the ink? We do it by firstly providing a tube and then ensuring that the ink has a specific viscosity to prevent flow out of the tube. This is depicted on the FAST diagram by 'provide tube' and 'control viscosity'. The rest of the diagram can be built up in the same way, firstly asking 'how' and then 'why' until the whole system is represented on one diagram.

Double checks can be made by asking the alternative question once a new function has been identified. For instance, having defined the function 'contain ink' by asking 'why' reveals that this is also required to 'prevent spillage' as well as to 'provide ink flow' at the ball.

A check now against the functions we wrote down at the start shows that all of those have been covered by the FAST diagram (except for function 6 which is optional). However, the FAST diagram has provided much more than a list of essential functions. It has provided the linkage between all of the functions of the system. If one looks at each end of the FAST diagram one can see that it has in fact linked the customer's requirements 'enable writing' with the essential components of the system. Extending the diagram more to the left will define further the elements of the components.

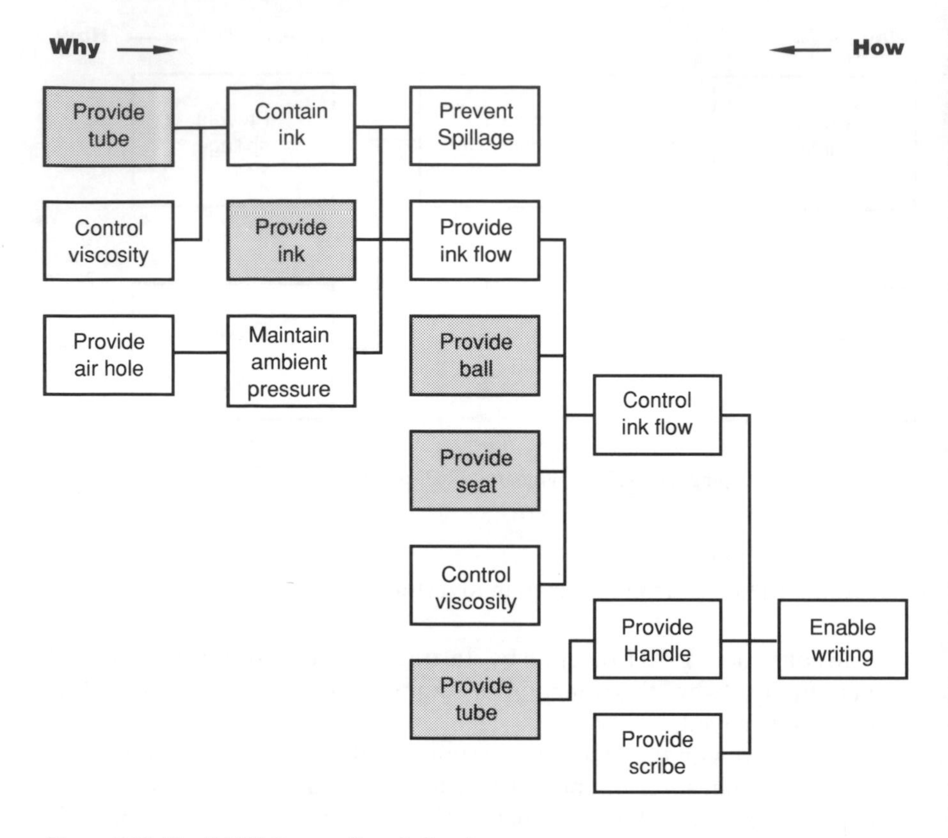

Figure 3.23 The FAST diagram for a ball-point pen.

What can we gain in terms of greater understanding and information from the FAST diagram? (See Fig. 3.23.) Let us first of all try to group some of the functional boxes. All those with the word 'provide' represent the provision of some part in the construction of the pen. (There is one exception to this, which is 'provide scribe'!) The parts list for the pen results in:

Tube (1) for ink
Air hole
Ink
Ball
Seat
Tube (2) for handle
(Two tubes are listed here, one to contain the ink and the other for use as a handle. Is there an opportunity to combine these parts into one which can do both jobs? In fact this is a real opportunity to value-engineer a device as simple as a pen.)

Secondly, let us look at all the other functional boxes and see if they represent any further understanding of the design. They are:

Control viscosity
Prevent spillage
Contain ink
Maintain ambient pressure
Control ink flow
Control viscosity
Provide metering gap

We can observe that control viscosity occurs twice, once to prevent the spillage of the ink and then to provide the flow of the ink. We can therefore conclude from this that an important aspect of the design of the system is to design an ink that has a viscosity satisfying two criteria. Are these criteria in contention with each other, that is can we satisfy both with one value of viscosity? The designer is immediately guided towards this question by the use and study of the FAST diagram. Furthermore, if the answer is obvious that both functions cannot be satisfied using one viscosity, the designer will be channelled to seek other solutions such as containing the ink in a different way. Here is an example of how a designer can obtain essential information about a major potential constraint on a system before he has even started to design in the conventional way and certainly before any hardware is built or tested.

There is an important requirement to provide a controlled gap between ball and seat in order to meter the ink. This tells the designer that there is an important interface between the ball and the seat that retains it and that this should receive close attention in terms of tolerancing, etc.

Finally, the designer may observe that the system requires that the pressure in the tube must be maintained at ambient to ensure adequate flow.

In summary, the FAST diagram in this case reveals the important aspects of the system for the designer to consider as the details of the design are worked out:

1. Flow of ink is important to eliminate spillage and to provide good performance.
2. The ball and seat are required to be closely controlled to enable metering of the ink.
3. Ink must be maintained at ambient pressure.
4. The system can be made using five parts.
5. There are opportunities to make two of these parts into one common part (tube for handle and containing ink).

This is a simple example of applying the processes of FAST to an already designed product. More usually the procedures are used to create the design of the system and then refine it and simplify it. The above example illustrates in simple terms the usefulness of the technique. It can be seen fairly easily that with a more complex system the FAST technique can be used to gain a clear understanding of how the system functions and to enable the designer to effectively reorganize the system design for improved utility of parts and reduced complexity.

The critical functions Considering the FAST diagram and the information generated from it, what are the critical functions of the system? Obviously, these are the design of the ink itself and the mechanical design to control the flow of the ink. Ink will be metered by maintaining a controlled gap between the ball and the seat in which it is housed and will flow according to this gap and the properties of the ink. Let us therefore list the parameters that will be needed to be kept in control in this case:

Ball diameter
Ball surface finish
Seat diameter
Viscosity of ink
Surface tension of ink

For this system these five parameters are seen to be essential to maintain good performance of the function. Each of these parameters will have a nominal value that is required to enable the function to occur. In addition to this, each of the parameters will have a range about the nominal which defines the tolerance of the parameter. These are the critical parameters of the function and are the only ones over which the designer has control. There are other influences on the function. The temperature of the ink will affect its viscosity. This temperature will vary with the ambient temperature and the heat from the user's hand. The user will apply varying amounts of force to the pen on the paper and this will cause the ball-to-seat gap to be varied, thus affecting the flow of ink. The angle that the user holds the pen to the plane of the paper could also have the same effect. These factors which are out of the control of the designer are called 'noise factors'. Each one of these has a range through which it can change, just like the critical parameters.

The critical parameters and the noise factors are the only parameters that enable the function to operate. Once this happens the designer needs to understand how the function can be assessed. What is the factor that can be measured that will assess how well the function is performing (Fig. 3.24)?

Let us now consider the primary function 'control flow (of ink)'. How would we measure whether flow control is being achieved? Obviously, by weighing the pen accurately before and after use to draw a line of a certain length, one could determine the flow rate of the ink.

Now consider what might happen if the flow rate of the ink were to increase to the point where the first failure occurred. Firstly, the amount of

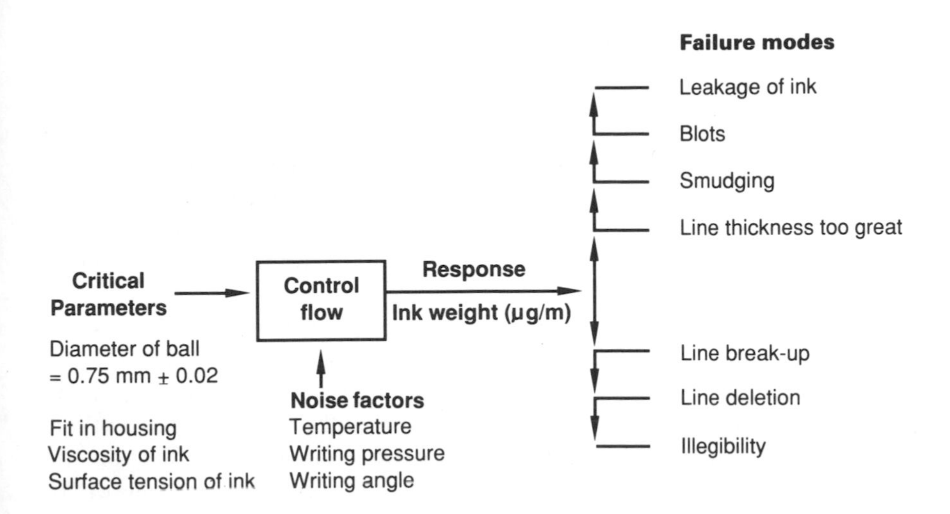

Figure 3.24 The function diagram for control of the ink flow.

ink dispensed to the paper would become greater and would cause the thickness of the line to increase. If the thickness of the line increased to a point where the customer was dissatisfied, perhaps because the fineness of the writing became destroyed, this could be considered as a condition denoting failure. In a similar way consider what might happen if the flow of ink gradually decreased. This might cause the line to reduce in thickness to the extent that it began to break up. This represents a failure at the other end of the scale.

The function, its measurable response (ink flow) and the failure modes representing flow that is too great and flow that is too small can be represented as part of a function diagram as shown. Of course the failure modes mentioned here are not the only possible failure modes. It is useful to layer the failure modes relating to increasing and decreasing values of the response. In this case further increase of the value of the response, ink flow, might result in smudging of the written text, blots of ink in the paper and finally leakage of ink, even when not being used to write with. Conversely, a reduction of the ink flow could cause intermittent line deletion and finally a complete inability to write.

The function diagram can now be drawn to include all the possible failure modes relating to the function of controlling the flow of ink. The difference in the values of the flow rates which represent the first failures of the function, that is maximum line thickness to line break-up, represents the latitude of the function and this latitude is unique to the specific design to which it applies. Also, the point at which a failure is said to occur is dependent on our interpretation of the customer's requirements. In other words, if we thought the customer could tolerate a greater thickness of line, this would mean that we would be thinking of a somewhat broader latitude for the same design. Of course the size of the latitude band is very important to us as designers. If the latitude is very narrow, we will be faced with designing a system that is very closely controlled in its output and this will inevitably lead to tighter tolerances in the design of components and the cost implications which that involves. If, on the other hand, the latitude is broad, we can expect a robust design to emerge from using conventional tolerances. This may lead us to explore whether the function can be designed in a slightly different way at less cost.

In terms of design it is important to understand exactly what factors will enable the control of the flow of ink. All the parameters that control this are not necessarily within the control of the designer. Of those that are, some will have a significant effect on the response from the function, while others will have a minimal effect. Those that have a significant effect on the function are called critical parameters and it is this set of parameters to which the designer must pay most attention. Let us attempt to list the critical parameters for the function of controlling the supply of the ink. Possible critical parameters are:

Diameter of ball
Fit in housing
Viscosity of ink
Surface tension of ink

For each of these, suitable units must be specified and nominal values and tolerance ranges assigned.

There are other parameters that could affect the function that are outside the control of the designer. These are called 'noise factors' and in this case could be such parameters as temperature. The designer must take these parameters into account in the design and ensure that whatever the range of values for the noise factors the response of the function will still stay within the range of latitude. The designer has only the critical parameters to enable this effect to be controlled.

With these elements added the function diagram can be completed. Once done, this gives the designer a full understanding of the function for which the design is being made and enables this perspective to be kept throughout the whole design process.

Using the FAST Diagram

The FAST diagram helps the engineer's understanding of the functional process in a number of ways. Two have been mentioned already, namely the ability to determine the breakdown of both cost and reliability. More generally, the FAST diagram helps the designer or engineer to obtain an overall picture of the functional relationships within the system. It may be obvious where there are areas that are not optimal in terms of the way the system is structured. If it is not obvious, a quick look at the diagram will help to focus on the areas that appear to be cumbersome or somewhat complicated for the functions that are being achieved. Judgement of this is obviously somewhat intuitive but after some practice using this technique the designer will develop a 'feel' for such areas.

FAST diagrams can be used once the design has been initiated to apply value engineering techniques which may be invaluable to developing competitive designs where cost is paramount in the engineering targets. (This is dealt with later in the book, see Sec. 3.6.) The FAST diagram also provides the gateway for the functional analysis of the system and the individual subsystem areas. Referring back to Sec. 3.4 under the function diagram, it is the individual boxes (or groups of boxes) from the FAST diagram that form the initial box of the function diagram. To see how these are related, let us now continue to use the example of the office stapler to develop the function diagrams from the FAST diagram.

In theory one could use each of the functions from the FAST diagram and form a function diagram from each one to fully represent the system. This of course is not practical and we must therefore decide which of the functions within the FAST diagram are key to the overall function of the system, bearing in mind that we are primarily interested in latitude and failure modes.

Failure modes can be defined either from the experience gained from testing the functional operation (in the laboratory or the field) or by analysing the operation and predicting failures using a technique such as failure modes and effects analysis. When a design is immature, few data exist on its performance, so here the engineer must rely largely on analysis or on knowledge gained from previous similar products. Where a system is either mature or is an evolution from an earlier product, the engineer can use data that will be available. In the case of our example here, we, as users, have a great deal of knowledge about the main failure modes of a desk-top stapler. Knowing this leads us to the most important functions contained in the FAST diagram.

Most people will I am sure agree that the most likely failure of the stapler is its failure to staple or to cause the staple to jam. Looking at the FAST diagram it would seem that this relates most closely to the functions 'drive staple' or 'clinch staple'. Using the FAST diagram now to search for other potential areas of failure, one might be drawn to 'accommodate staples' or 'open throat' as the next most likely failure areas. If indeed these were of concern to the designer the choice could be made to develop and study a function diagram for these.

For the purposes of this example, however, we will develop the function diagram for 'drive staple' as our data tells us that this is the key failure area. The function diagram starts by drawing the function box and inserting the words 'drive staple'. The next step is to ask what response we expect from this function. In other words, what will we measure to confirm that this function 'drive staple' has actually happened? The measurement of this can be thought of in laboratory terms, that is the use of special instrumentation may be needed to accomplish the measurement. Concluding this question, what will we measure to confirm that the function has been accomplished, will enable us firstly to develop more accurately the failure modes and under what circumstances they occur, but, more importantly, will ensure that when we want to test this function for reliability we actually test and measure the correct factor.

What can we monitor or measure to ensure the function is accomplished? Ideally we want to monitor the position of the legs of the staple as the stapling operation proceeds. The response therefore to this function is 'position of staple leg'. This can now be added to the function diagram as a response as shown in Fig. 3.25.

Let us analyse carefully the failure modes that could occur from this. Referring to Fig. 3.26, it can be seen that the response is assessed by measuring the distance that the staple leg has penetrated the paper stack. The dimension used is the distance from the end of the staple leg to the underside of the stack. Negative values of this will mean that the staple has not penetrated right through the stack of paper, whereas positive values describe the extent to which penetration through the stack has occurred. No consideration is given to the clinching of the staple after it has been driven through the paper stack. As can be seen from the diagrams, a satisfactory penetration by the staple is represented by figures in the range of 2 to 4 mm. Negative values represent either a mis-staple or a partial staple as shown. Positive values above 4 mm represent an over-staple which has caused the top of the staple to damage the top of the paper stack.

The 'layering' of these failure modes is depicted in the function diagram.

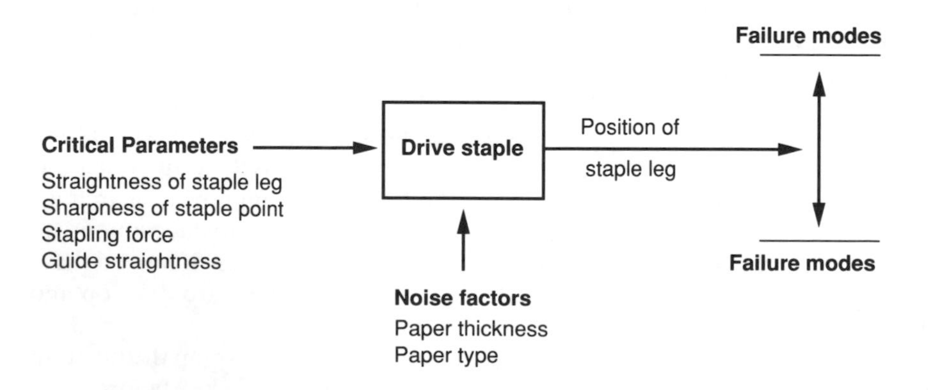

Figure 3.25 The function diagram for 'drive staple'.

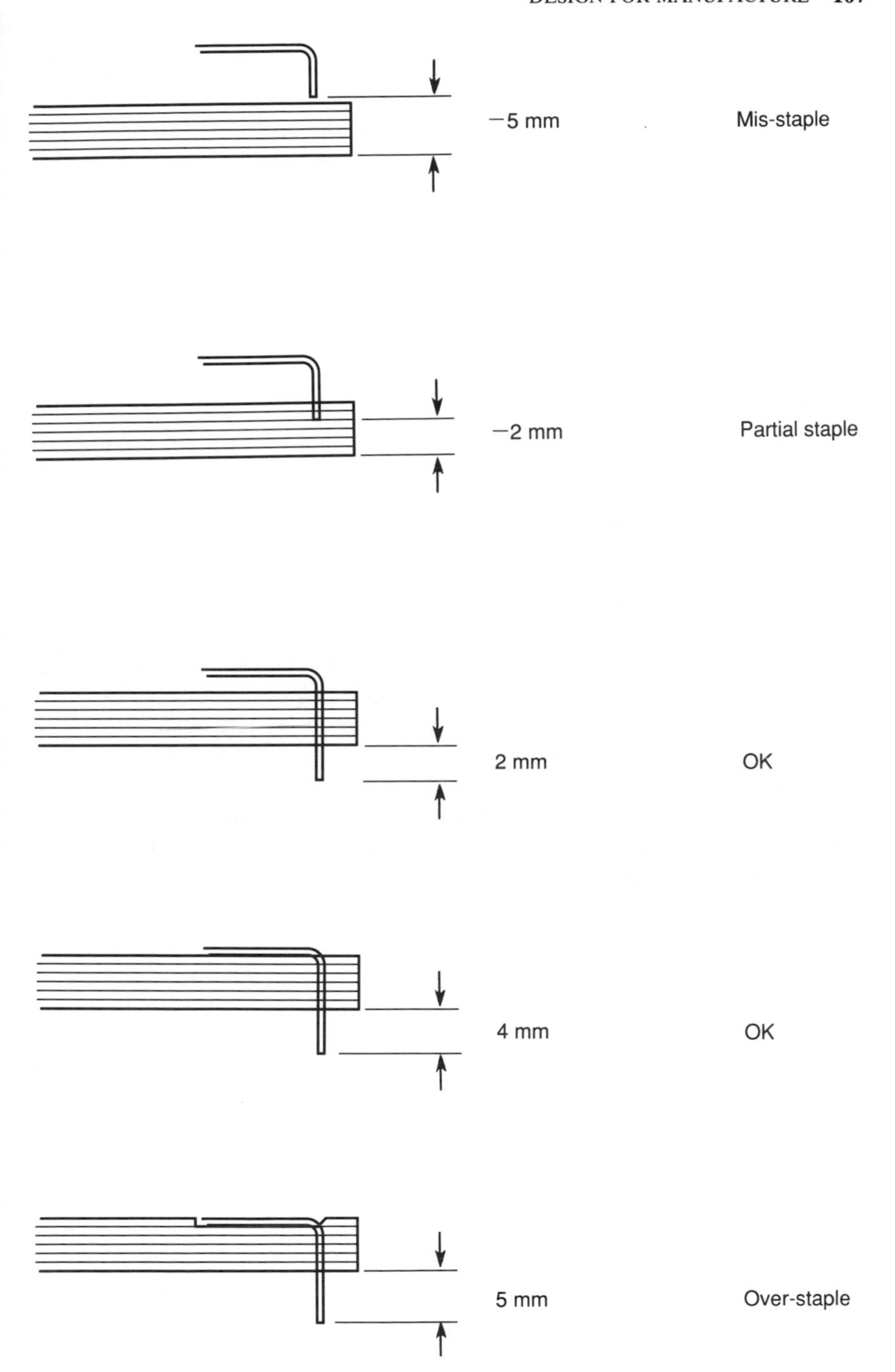

Figure 3.26 Stapling failure modes.

In this way the extremes of the satisfactory movement of the staple can be delineated. The range of these values which can occur without causing a failure, that is from +2 to +4 mm, represents the latitude of the system.

Now consider the parameters that control the function of 'drive staple'. Thinking about the action of driving the staple, it seems evident that the critical parameters are:

Straightness of the staple leg*
Sharpness of the staple point*
Drive force
Straightness of the guide
Clamping force of paper stack

There are some other factors that affect the function which are noise factors. These are:

Paper stack thickness
Paper type

It should be noted here that the critical parameters marked * could also be noise factors if one considers that the staples themselves are outside the control of the designer. This could well be the case when users do not employ the recommended staples for the device.

The designer is limited in control of this function. The designer's job is to make the device as reliable as possible by creating a design that uses only the critical parameters to control the operation, in full knowledge of the noise factors that prevail. Robustness of the design can be achieved by considering the three critical parameters which are independent of the staples being used, namely:

The drive force
The straightness of the guide
The clamping force on the paper stack

The level of each of these parameters must be assessed according to how it affects the response 'position of staple leg'. Too high a drive force, for instance, may cause over-stapling; too low a drive force may cause mis-stapling. Obviously, high accuracy in the straightness of the guide will give a better performance, but how much tolerance can be allowed for this to remain satisfactory? Is the clamping force a critical parameter? How does it affect the position of the staple leg after stapling? Once these relationships have been established the designer can recommend nominal and tolerance ranges for each. The optimization of these values, including any interaction that may be present, can be determined by experimentation and the designer can refine the parameter values to meet the desired performance. The success with which this is achieved will define the robustness of the design. Lack of robustness will mean that the design will be more likely to be subjected to failure as a result of its inability to compensate for the noise factors. A robust design will, on the other hand, provide reliability and consequent customer satisfaction will result.

Using FAST to Develop Competitive Products

As we have seen, the FAST diagram is a convenient way of representing the functional relationships of the system. The words in the boxes are used to denote the functional elements of the system and the arrangement of the boxes is used to represent the way each function relates to the next and the way the system fits together. Each of the boxes represents more than just the functional element. After all, we can enter any parameter associated with that particular part of the system and see what its relationship is with the others. This is true because having laid out the total network this is a valid representation of the system in all its aspects.

Two of the most important of the parameters in which we are most interested when considering the development of a competitive product are cost and reliability. The FAST diagram can be used to provide an analysis of both these factors and enable us to compare either with competitive products already on the market or to develop a highly competitive product by looking at a detailed breakdown.

Cost Once the FAST diagram has been drawn and optimized as a system, each functional element can be considered from the point of view of cost. For each of the boxes, the cost of manufacturing the parts that contribute to the function is estimated and written into the box. For further refinement a split between materials and labour costs can be entered separately. However, for the moment we will just consider the overall cost. Once a group of boxes has been costed separately, the sum of these costs can be entered into the box which is fed by the individual functions. In other words, costs accumulate as one moves to the right of the diagram. Of course, it is essential that no double counting occurs. As each of the boxes is allocated a cost and the cumulative values are compiled, the box at the right-hand side of the diagram represents the cost of providing the full functional system or the total product. This of course gives an indication as to whether the arrangement of the functional system has been optimized to a level sufficient to provide a viable product.

Moving now to the individual boxes that make up the diagram, these give the designer a feel in each case as to whether the particular function represented by the box is a reasonable figure and whether it gives reasonable value for money. As we move to the far left of the diagram, the boxes represent individual parts and it is here where the basis for the costing occurs and where it can be modified if necessary.

A FAST diagram can of course be drawn quite easily for a competitor's product by dismantling the product and understanding the functional operation and relationships of the individual parts. Then cost can be entered into the boxes based on our own estimates of the competitor's parts. With a competitor's diagram to compare with our own, we now have a valuable vehicle for seeing where the strengths and weaknesses are in both cases.

Firstly, the overall costs can be compared. Secondly, we can look for similar functional areas and compare the costs of achieving this for the two designs. Finally, we can see where parts have been made more cost effectively and decide whether this is a function of the design itself or related to the design of the system. Whichever is the case, we can take steps to build on the strengths of our own ideas and see whether competitive ideas would help us to further improve the overall design.

Let us now look at an example of how cost information can be examined using a FAST diagram. Figure 3.27 shows a detailed FAST diagram for the desk-top stapler shown in Fig. 3.18. Figure 3.28 shows the parts list with cost estimates for each of the 27 parts. The costings of the parts include the material, the manufacturing cost and the finishing, such as plating or anodizing. An assembly time of 4.8 minutes adds £1.36 to the overall cost, giving a total of £5.82. If we look at the main cost drivers in the assembly we see that they are the base (01), the body (19) and the spindle (21). There are some rather expensive screws in the assembly (13, 15 and 24).

We will see later how, by value engineering the design, cost can be reduced both by reducing the number of parts used and by combining functions.

Reliability Reliability can be treated in a similar way. However, where in the case of cost analysis we started at the left-hand side of the diagram, when considering reliability we should start at the right-hand side.

The designer should be aiming for the lowest levels of failure consistent with the reliability target for the product. In the same way that we dealt with cost, allocate reliability across the functions that 'feed' it. Work towards the left of the diagram, gradually breaking down the reliability and allocating it to the lower level functions.

With a diagram depicting the full allocation of reliability across the product, we can start to analyse and optimize a number of aspects. What, for instance, are the major areas driving low reliability? Which are the areas with comparatively high reliability? How do these compare with the equivalent areas of the competitor's product. In particular, looking at the FAST analysis of cost, are there areas that could possibly be made more reliable by increasing the cost? Conversely, are there areas that could safely be cost reduced without danger of reducing the reliability to an unacceptable level?

A typical failure rate of a desk-top stapler is about 50 failures per million operations. This means that when the customer staples the set of pages there will be dissatisfaction either with the failure to staple or the quality of the

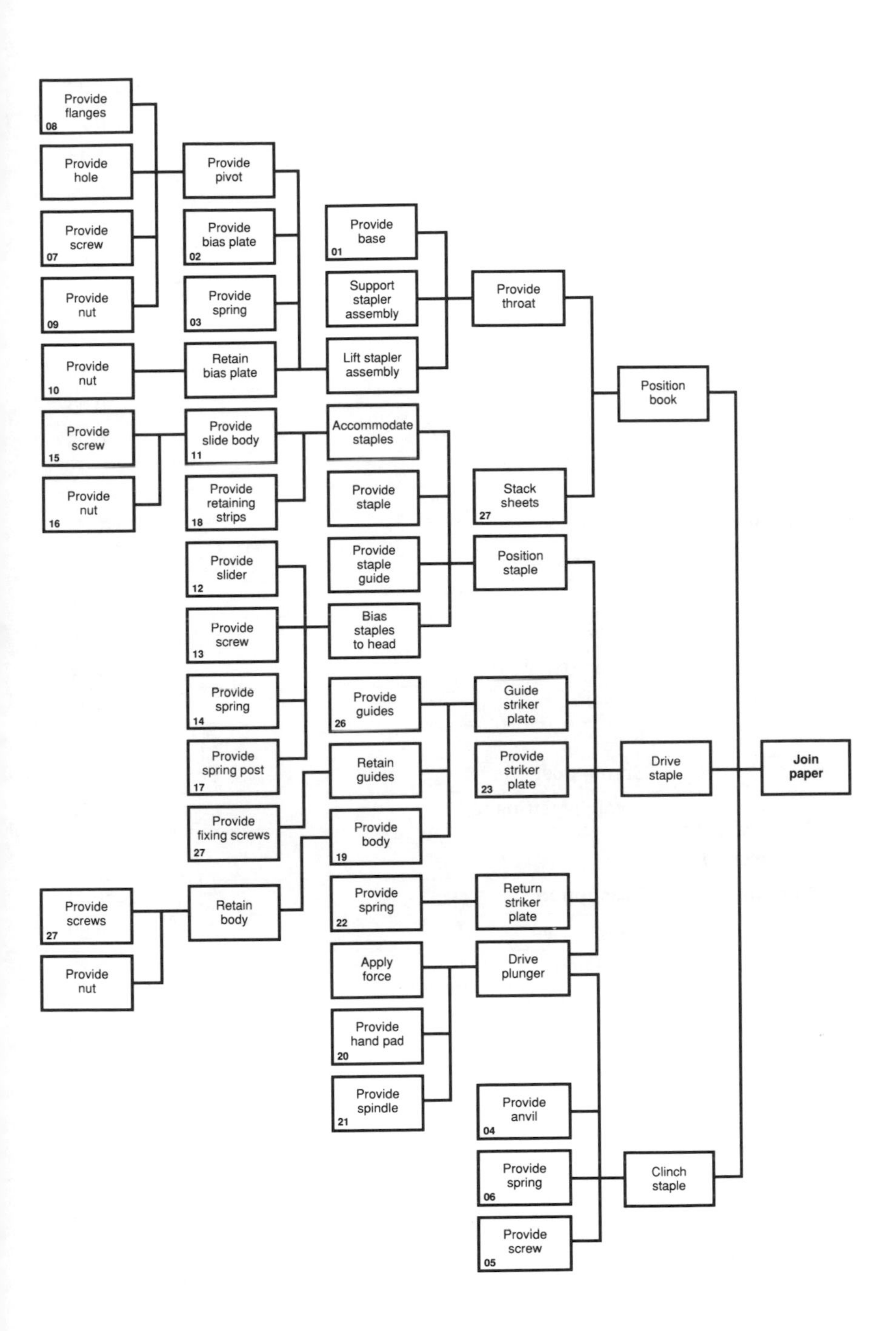

Figure 3.27 FAST diagram for desk-top stapler (in detail).

Base assembly

Part number	Description	Cost (£)
01	Base	0.75
02	Bias plate	0.15
03	Spring (compression)	0.09
04	Anvil	0.15
05	Screw	0.06
06	Spring (compression)	0.05
07	Pivot pin	0.15
08	Flanges (2)	0.12
09	Nut	0.06
10	Nut	0.06

Slider assembly

Part number	Description	Cost (£)
11	Slide assembly	
12	Slider	0.12
13	Screw	0.18
14	Spring (tension)	0.12
15	Screw	0.21
16	Nut	0.06
17	Spring post	0.18
18	Retaining strips (2)	0.06

Stapler assembly

Part number	Description	Cost (£)
19	Body	0.60
20	Hand pad	0.25
21	Spindle	0.30
22	Spring (compression)	0.09
23	Striker plate	0.08
24	Screw	0.21
25	Nut	0.06
26	Guides	0.18
27	Fixing screws (4)	0.12

Assemble (4.8 min) = £1.36
Total = £5.82

Figure 3.28 Parts list and costing.

finished result 50 per 10^6 times. Sometimes the failure will be a failure of the mechanism to operate correctly. At other times it will be a failure of the staple itself to adequately join the sheets and finally there will be times when the operator fails to make a good job of the operation, say by not aligning the sheets in the set. All of these conditions can be addressed in some way or other by the design. Some will require more cost to be added to the device. Others will be implemented at the same cost but will need more innovative design.

Looking at the FAST diagram (Fig. 3.29), the failure rates for many of the high level functions have been added. Starting with the final function 'join paper', this shows the overall failure rate of $50/10^6$ operations. This rate has been apportioned, either from test data or just from judgement, across the three functions which enable 'join paper'. The numbers say that the largest contributor to the failure rate is the function of driving the staple ($40/10^6$). Next come the clinching of the staple and the positioning of the book by the operator (both at $5/10^6$). This fits with our own experience of these devices, where by far the greatest source of failure is actually driving the staple from the staple supply and getting it successfully through the thickness of sheets.

Let us look more closely at what goes to make up the failure rate of the driving of the staple. The further breakdown of failures shows that the contributors are the positioning of the staple under the striker plate ($15/10^6$), guiding the striker plate itself ($12/10^6$), driving the plunger divided equally between 'drive staple' and 'clinch staple' ($10/10^6$) and getting the striker plate to return to its home position ready for the next operation ($8/10^6$). With this information on the allocation of failures we can now begin to look at where there are shortfalls in the design and start to search for what might be done about them. For instance, if we look to the left at the functions that support 'position staple' we see that there is a guide for the staples. Could this guide sometimes prevent the correct positioning of the stapler prior to driving? Also there is a need to bias the staple up to the head. Perhaps this bias is sometimes inadequate and causes a malfunction. We can therefore begin to look further to the left to see how we have designed specifically for these requirements and start to ask if these designs are adequate or whether they could be improved. For example, the spring to provide the bias may be too weak or the guide and the staple being used may not be a compatible fit.

Moving on to guiding the striker plate, it needs a good fit between the guides and the body. Is this fit acceptable? Could it be affected by temperature changes or dirt? If so, how could this be prevented?

Tracing the functional relationships through the FAST diagram and understanding the apportionment of failure rate is a major enabler to finding shortfalls in the design and searching for solutions. When compared with competitive products the same allocations of failure rate can be made and

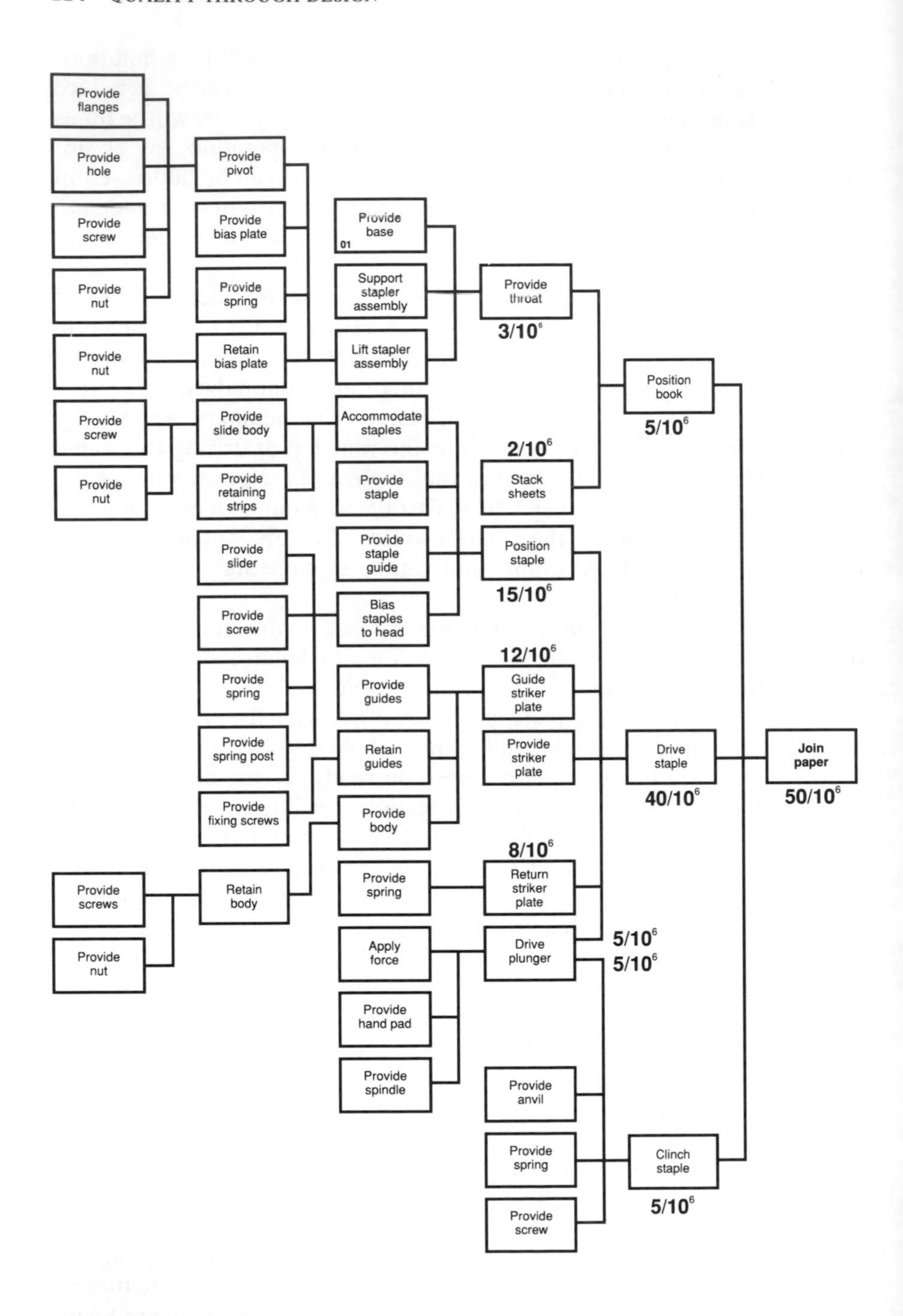

Figure 3.29 Reliability analysis through FAST.

the designs that appear to have a lower rate in the competitive products can be compared with our own design to isolate the areas of weakness. The process also leads the designer in the direction of trying to eliminate unreliable areas of the design. For instance, the plunger contributes to two areas of failure, driving the staple and clinching the staple. If some of the parts in this area could be eliminated this may provide gains in reliability in more than one part of the product. In this way the designer might pursue the avenue that modifies the design to reduce the parts count in this area.

The Use of Critical Parameters

The preparation of the function diagram forms the heart of the process of design. The picture that it reveals for the design team is rich in information which will guide the design to a successful conclusion and deliver a quality product. The list of critical parameters forms what is described as the 'design intent', that is the designer's best attempt at any stage to actually define the elements of the design. Of course the full design is described by the complete set of parameters, not just those that are critical. As we have seen before, the design is controlled by the critical parameters, so by attending to these the designer essentially retains control of the full design. Defining the design intent in the form of a list of critical parameters, each with its nominal value and range, gives the designer the full input information to embark upon the design layouts and files. One of the problems in the past that has contributed to poor design has been the fact that it has been assumed that the drawings when completed have been a true reflection of the design intent and, moreover, were accurate in terms of tolerance build-up and fit. Additionally, when the hardware prepared from these drawings was received, the question of inspection of parts arose and whether 100 per cent inspection was required or, if not, which dimensions should be checked. Often the effectiveness of this process was not revealed until the hardware was assembled or even tested.

Using the design intent as defined by the list of critical parameters allows a much more rational approach to monitoring the design throughout the whole process. This is called the critical parameter tracking process.

Critical parameter tracking process This process uses critical parameters to monitor design quality through drawing and the building of hardware (Fig. 3.30). Once the drawings have been completed and before they are issued for build the designer is required as part of the programme management process to calculate for each of the critical parameters the range of values that will be delivered by the drawings that have been completed. A direct comparison between this range and the value of the design intent for this critical parameter will show immediately whether or not the drawing meets the design intent. Normally, of course, any deviation would be

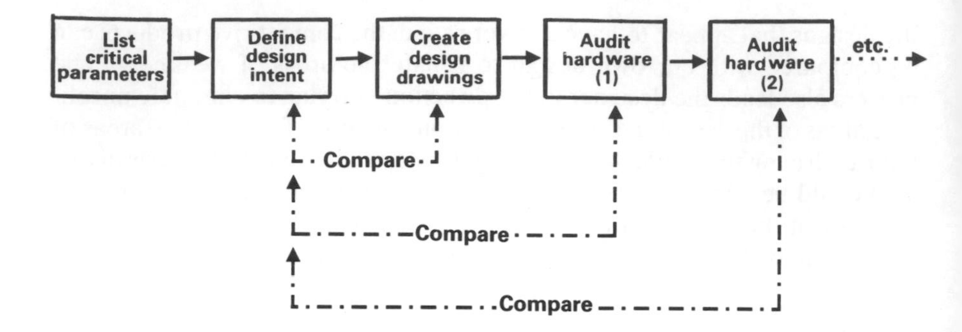

Figure 3.30 Critical parameter tracking process.

corrected by making the appropriate changes to the drawings. However, this comparison need not necessarily be used to create a go/no-go situation, because if the deviation of the drawing from the range of the design intent is known, then the effect on the function in terms of latitude and failure mode is also known. Rather than rejecting a drawing status that is outside limits, the chief engineer can, with this knowledge, choose how to proceed. It will be known, for instance, what sort of failure mode the deviation is likely to introduce, and this can be weighed against the cost and schedule risks of making the correction at this time. The engineer may decide to accept the shortfall and plan to make the correction at a later stage in the programme. The important point is that the risk is able to be managed based on sound information about the functional shortfall of the design.

Comparing the drawing status with the design intent for each critical parameter provides a valuable check of the drawings prior to issuing them for manufacture. One may argue that checking only critical parameters may allow a drawing to be issued that will produce a part that is incorrect. This is true, but any parameter that is not on the critical list has, by definition, wide latitude of function and is robust against noises that the system may undergo. It may, however, suffer a fit problem and this of course will need to be corrected. By and large fit problems can be resolved fairly easily on the hardware and although everything that can be done to eliminate these errors should be done, if the hardware is free of functional problems, this has to be a major step forward.

Hardware, unless it is production hardware, is built for test purposes of some description. Critical parameter tracking is used to validate the hardware prior to testing. As parts are received and assembled into the system, the actual values of the critical parameters are audited and compared with the design intent. This serves two purposes. Firstly, the quality of the hardware is monitored against the design intent and, secondly, a record is made of the critical parameters at the start of the test. This last point is important because so often a test will start and not until there is a failure will the important parameter be measured. By then nobody knows what the parameter was to start with so a valuable piece of data is lost and the test will not be able to identify whether or not a condition developed to cause a failure.

The basis of the critical parameter tracking process is to compare every embodiment of the design with the design intent. Deviations from the design intent may be corrected or allowed to prevail, depending on the level of understanding of the effect on performance.

A device used in copying machines known as a friction retard feeder provides a good example of the use of critical parameters to simplify the function. The device is used to separate sheets of paper in a stack and feed them into the transport system of the copying machine. It works on a similar principle to the way we often separate sheets of paper using our fingers. We normally try to break the frictional forces between sheets by using our forefinger and thumb to slide the uppermost sheets in the opposite direction from the lower sheets. The success of the operation depends very much on the relative frictional levels between finger and paper, thumb and paper and between adjacent sheets. We often modify these by wetting our fingers to improve the friction between our fingers and the sheets of paper.

Research and development work has been able to analyse the physics of the function of this device and one of the early embodiments was configured as shown in Fig. 3.31. The device consisted of an elastomeric belt driven around two pulleys with a friction pad penetrating the assembly between the pulleys causing a deflection of the belt. Tension in the belt was controlled by the centre distance between the pulleys. The device operated by the application of a force between the paper stack and the left-hand pulley. This pulley

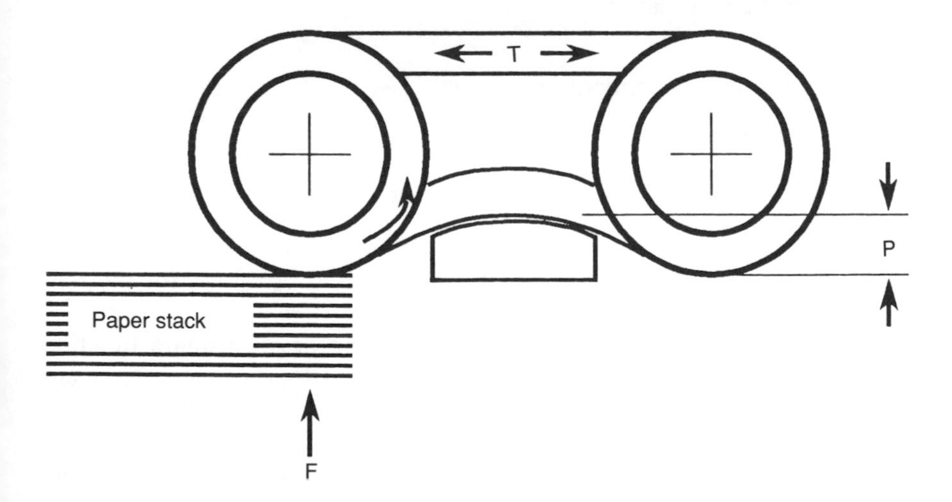

Figure 3.31 A friction retard paper feeder.

rotating anticlockwise caused sheets to be driven into the space between the pad and the belt. (The diagram is only a schematic of the geometry in this area.) The pad-to-paper friction and the belt-to-paper friction, being higher than the paper-to-paper friction, caused sheets to be separated from each other. The separation effect increased as the sheets progressed over the pad, until only one sheet emerged at the exit from the pad. Thus the device separated sheets from a stack and enabled single sheets to be introduced into the paper path of the copying machine. Following the selection of materials to achieve the correct frictional values, the three parameters that were considered to be of paramount importance to the function were:

1. Stack normal force (F)
2. Belt tension (T)
3. Pad penetration (P)

The design went ahead using these as the controlling factors and the device eventually became part of a product that was manufactured in large quantities.

If one studies these three parameters and considers the difficulties in controlling them through the design and manufacturing, it becomes clear that the design really addressed the wrong parameters. Belt tension is dependent on:

Belt length
Belt thickness
Pulley centre distance
Pulley diameter
Pulley run-out
'E' for the material
Belt wear

Pad penetration is dependent on:

Pad thickness
Pad wear
Pad positioning

In order to control the original three parameters all of the above must be controlled through design, manufacturing and field operation for the reliability of the device to remain high.

If we consider the critical parameters of the device using a function diagram, we obtain Fig. 3.32. The diagram shows that instead of concentrating on belt tension and pad penetration, the true critical parameter for the belt–pad arrangement is the force between the pad and the belt. Giving this further consideration in terms of the design produced a dramatically improved version of the design which was much less dependent on the production processes and the wear in the field throughout the product life.

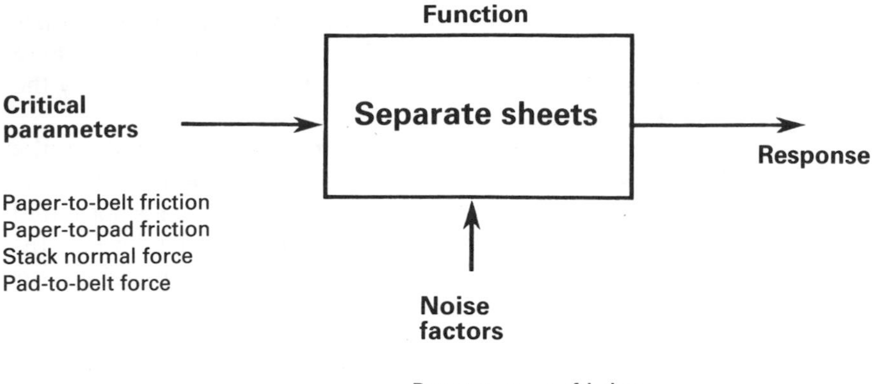

Figure 3.32 The function diagram for the retard feeder.

In addition to this the ultimate design turned out to have less parts and was therefore cheaper.

The diagram of the improved design (Fig. 3.33) shows a single elastomer roller driving the sheets off the stack and a single spring-loaded pad acting as the retarding component. The force between the pad and the drive roller is the critical parameter. It is independent of roller diameter and run-out variations. It is also independent of wear from both the roller and the pad. As a result, the design is more robust, and at the same time is cheaper due to

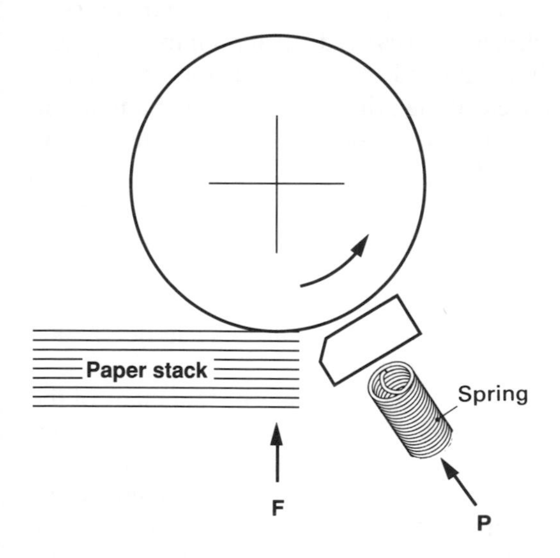

Figure 3.33 An improved design for a retard feeder.

the reduced number of parts and the elimination of the need for settings during manufacturing. These benefits flow through to the field where there will be no need to reset the device to compensate for wear; generally the device will be more reliable.

The lesson to be learned from this is that, firstly, the attention of the designer should be focused on the fundamental critical parameters of the function, not some derivative of them. Secondly, a design that is optimized in this way is rarely more costly. On the contrary, it often contains benefits that reduce not just the base manufacturing costs but also the operating costs of the device due to greater reliability (less servicing) and longer life.

3.5 CRITICAL PARAMETER MANAGEMENT

The approach to the design process using critical parameters also has enormous benefits in managing the subsequent design. Designs have to be managed throughout the life cycle of the product. This means managing the design from preconcept, when ideas are just being formulated, through the creation of the design by drawings or design files, through manufacturing, through the launch process and into the field.

We have seen how the design can be tracked by monitoring critical parameters in the drawings and by checking them in the parts. Test hardware can be audited prior to test by checking the absolute critical parameters and then when a failure occurs any deviation from the norm can be identified again by rechecking the critical parameters.

Once the product moves into manufacturing, any deviation that is seen in the system performance, which is usually carried out at a number of points during assembly and also finally at the end of the production line, can be readily checked by a simple recheck of the critical parameters. Similarly in the field, any deviation in performance can be identified in the same way and steps taken to correct it. This is made even more effective if the diagnostic processes are written around the critical parameters in the first place. In short, the critical parameters, since they represent the enablers to the function, are the quickest route to any malfunction that occurs and can be used in this way throughout the life cycle of the product.

The process of the management of critical parameters can be divided into three distinct parts for the design activity (Fig. 3.34).

When a concept or idea is initially conceived, it may be that many of the critical parameters are not known. Some of course may be known either from previous similar designs or perhaps from an analysis of the functional system. Others may be known simply from exercising good engineering judgement or general experimentation with the idea. Whatever the situation at this stage there will be a need to develop the known critical parameters into a comprehensive set whose general sensitivities to the response of the

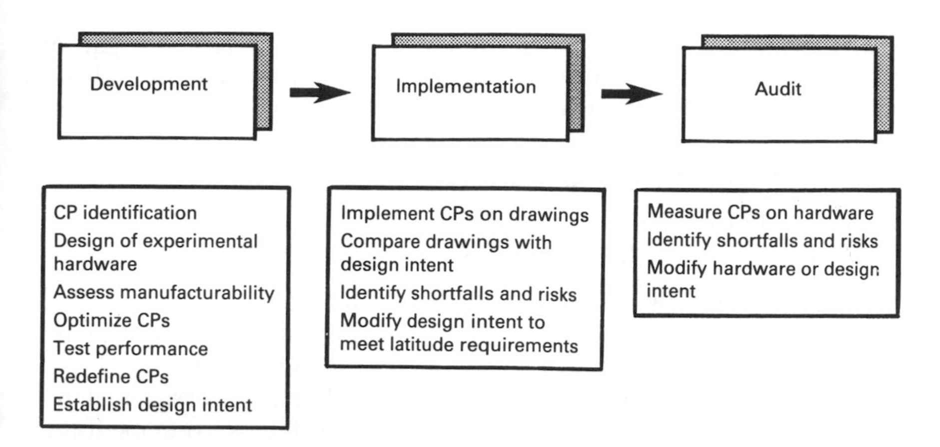

Figure 3.34 Critical parameter management.

function are known. During this process optimization of the assumed critical parameters will take place and the process should also identify shortfalls in the list of critical parameters and indicate possible additional parameters that are also critical and should therefore be considered. This part of critical parameter management is called critical parameter development.

Once the critical parameters have been used to define the design intent, they can be used to prepare the drawings or design files which will be the vehicle for the manufacturing of the parts. This is the critical parameter implementation process. It involves the confirmation that the drawings meet the design intent by monitoring the critical parameters. Once this has been done, the drawings may be released for build. The process enables a review of the design intent status and also makes a check on whether the technology selection is capable of meeting the design intent.

The third step of critical parameter management is the critical parameter audit process. Here the parts will have been manufactured following the release of the design drawings. The process enables a check to be made on the parts to identify any shortfalls against the design intent and to identify whether the parts are correct according to the drawing or whether the drawing itself is wrong. The focus here is again on the design intent. With this as a focal point the designer is prevented from falling into the trap of assuming that the drawing is the design. As we have seen, the drawing is not the design but only a representation of it and can sometimes be wrong.

Therefore the three elements of critical parameter management are:

Critical parameter development
Critical parameter implementation
Critical parameter audit

Once the product has reached the stage of manufacturing, similar processes to those described here can be followed to identify shortfalls in that part of the product life cycle. Servicing may also be improved by similarly addressing critical parameters in the diagnostic processes.

Critical Parameter Development Process

The detailed process of critical parameter development is depicted in the flow chart of Fig. 3.35. Specifically the process is as follows:

1. Having defined the function to be designed and developed the function diagram, the critical parameters are listed. Here quality and completeness of the list will depend on the level of maturity of the particular design being attempted. If the design is an evolution of a design that has gone before then it is likely that a great deal will be known about the function diagram and the critical parameters. If the design is an embodiment of a new idea or invention then both the function diagram and the critical parameters will be based on the designer's engineering judgement, level of previous experimentation and background analysis and research. It does not matter at this stage in the process if some of the fundamental critical parameters are omitted. Of course, if the list can be close to complete, then so much the better. The process, as will be seen, is tolerant to an incomplete understanding of the function at this stage. The critical parameters must be listed based on whatever knowledge is available from history, testing, analysis or any other means.
2. The next step in the process is to assign nominal values to each of the critical parameters listed. For this step almost all of the comments stated above apply, that is completeness is desirable but not essential. Most of the time the engineer's knowledge will be capable of making a very good judgement about this. Thinking about how the undisciplined process of design often works, the designer is expected to (is left alone to!) come up with the best judgement for all of the initial parameters of the design. It is usually the process of iteration during the following building and testing of hardware that promotes the changes to these parameters. In the disciplined process an initial best judgement of the nominal values is required and this can be modified in a controlled way as greater knowledge is developed through analysis, testing and optimization.

   Fundamental in assigning the nominal value is also to clearly define the units. Although this may seem an obvious statement to make, attention to this is a safeguard against error in this stage.
3. Against each critical parameter and its adjudged nominal value is added the adjudged tolerance or the range over which this nominal can shift. Tolerance here takes its literal meaning—that which can be tolerated to enable the function. Often these tolerances will stem from historical data on previous, similar designs but often the tolerance is based on what can be achieved from the relevant manufacturing process. Beware

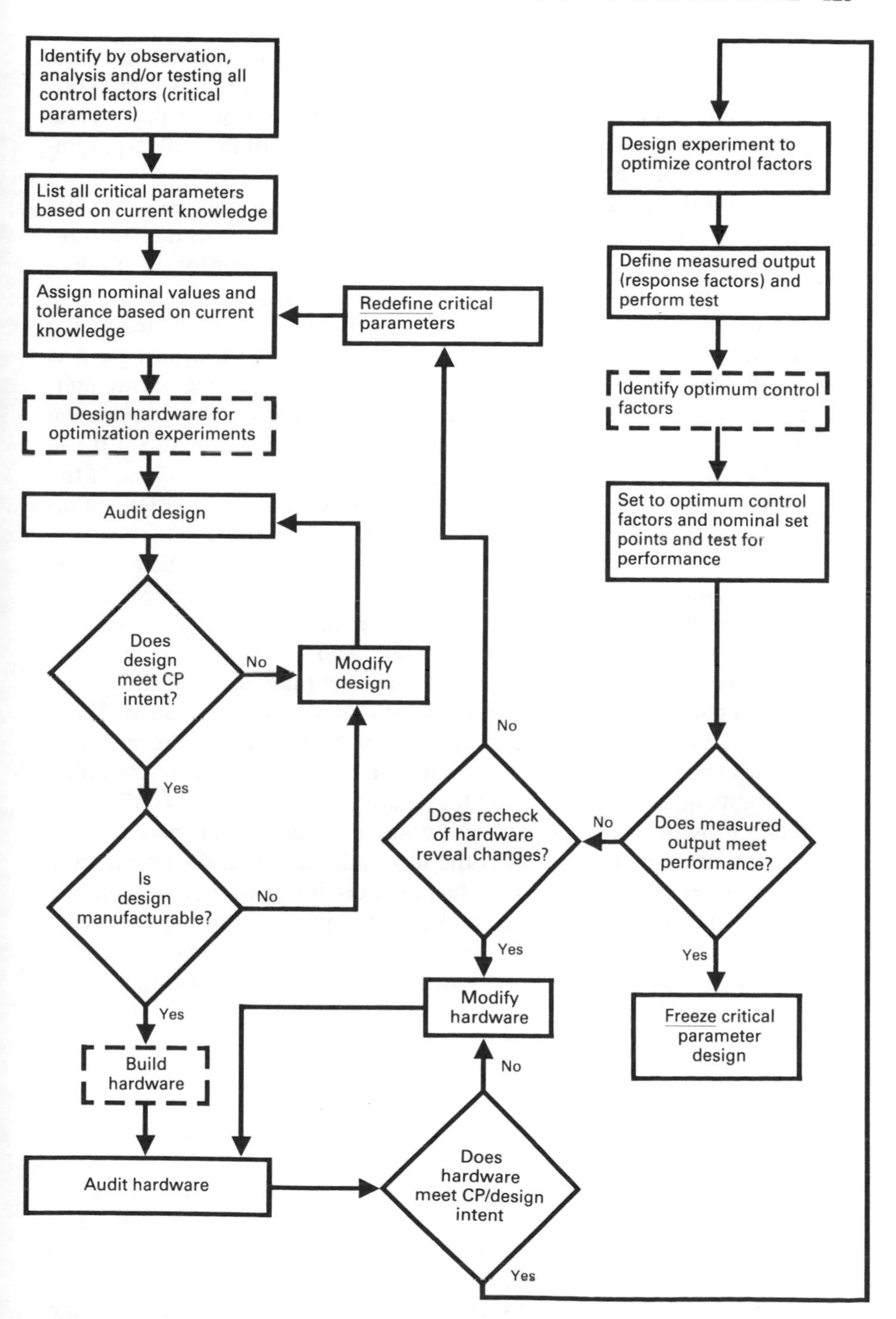

Figure 3.35 Critical parameter development process.

of this as it is not the best method of deciding the appropriate tolerance. Rarely will the tolerance be known to be critical to the function at this stage, but when this is the case the design engineer must pay careful attention to the sensitivities between the variation of the critical parameter, the response and the failure modes.

4. Based on these numerical data and other key information on product goals the design can be initiated either for the experimental hardware which will be used to develop the critical parameters or for the production design intent of the product. The basic process is the same for both. Once the design is completed, usually in the form of a design file (layout) or more specifically in the form of detailed drawings, it is audited by checking the level to which the critical parameters are met. For each of the critical parameters, the question is asked, 'Does the design meet the intent as defined by the critical parameters?' If the answer is 'No', then the design must be modified until it does. If the answer is 'Yes', the next question to be asked is 'Is the design manufacturable?' Here it is necessary to establish whether there is any parameter within the design that stretches the capabilities of the corresponding manufacturing process. It is probably not necessary to become involved in detailed discussion with manufacturing engineers or experts at this stage, since the engineer's knowledge of manufacturing processes should be great enough to highlight any areas of the design that might be close to the limits of manufacturing capability. These areas, if they exist, can form the subject for discussion with the manufacturing experts in the search for a solution. If the answer to the manufacturability question is 'Yes', the hardware can be built with confidence.
5. One of the problems of hardware going into test is that there is often such an urgency to actually get the test under way that the engineering team forget to check (or do not have the discipline to check) what they have in terms of parametric quality. The next step in the process is therefore to audit the hardware before testing in the same way as the design was audited prior to manufacturing the parts. Each of the critical parameter values is measured and recorded and compared with the design intent. The question is asked, 'Does the hardware meet the design intent?' If the answer is 'No', the hardware is modified until the question can be answered positively. Once the hardware meets the design intent, it can be used to embark upon the experiments that will enable the team to optimize the critical parameters.
6. The experimental work required to optimize the design of the function must of course be planned by the careful design of experiments. The design of these experiments inevitably has a significant effect on the design of the hardware itself, so it is likely and desirable that the original concept design of the hardware should embrace the objectives of the experimental work. The design of the actual experiments is often based

on Taguchi methodology, which enables the relevant sensitivities of parameters to be established.

7. It is an essential part of the preparation of this test work to define clearly the responses that will be measured to evaluate the testing. Guidance for this can again be derived from the function diagram, which will help the engineer to establish the most relevant responses and to design the instrumentation to measure them.
8. Conducting the experiments will identify the optimum levels of the critical parameters for the effective function of the design. It will also focus attention on whether there is another parameter that is critical but has not been included in the experimental evaluation. This will be observed by the fact that the responses which are being measured to confirm control of the function are erratic. Whenever there is a critical parameter that is not being controlled it will generally be manifested by inconsistency in the experimental results, particularly in the changes in a response. This should be a signal to the designer that one or more of the critical parameters is missing or out of control. The optimum values of the critical parameters will be derived from the experimental results and recorded as the first iteration of these.
9. The hardware must now be set up to reflect these optimum values and a further confirmatory test run and responses measured to establish functional performance. Once this has been completed and the results analysed, the question is asked, 'Do the measured outputs (responses) meet the expected performance?' If the answer is 'Yes' then the critical parameters are frozen and the design is optimized. It is more likely that the answer will be 'No'. At this point a recheck of the hardware must be carried out to ensure that none of the critical parameters has drifted from the original audit in item 5. If parameters are seen to have changed, then they must be reset and the experiments for optimization must be run again. Under these circumstances it is important to study why these parameters might have changed. Can they be reset as they were originally? Has some functional effect previously unaccounted for caused a change in one or more of the parameters? The need for a reset of parameters must be regarded as a warning to the designer that something may be wrong.
10. Assuming that the measured outputs do not meet the expected performance, but that parameters are all as they were in the original audit, there is a need to refine the critical parameter set either in terms of content or in terms of value or both. This demands a redefinition of the critical parameters, and the process reverts to item 3. What should be changed on the second iteration is something only the design team can judge, based on the knowledge gained from the first design exercise and the optimization experiments. Careful analysis of the results will almost certainly reveal other parameters or a fundamental error in either the

design concept or the assumptions made regarding one or more of the values assigned to the parameters.

This process model serves as an example of how one should develop the optimal critical parameters in a structured way. Inevitably it can be modified depending on the circumstances of the particular field of engineering. The process even allows the designer at this very early stage in the engineering process to make an early evaluation of the manufacturability of the design.

Critical Parameter Implementation Process

This process describes the methodology for the management of the production design intent (Fig. 3.36). This is manifested as a design layout for the design of the final product at whatever stage the progress of the design is in. That is to say, the design team, having embarked upon the production design for the product, will need to be sure that the critical parameters are tracked and controlled throughout the development of this design. The design will come under pressures from vendors who will supply the parts and will therefore want some say in how the design is done. It will come under cost scrutiny from management with pressure to reduce numbers of parts and reduce their costs by design modifications. There will also be many other pressures on the designer from outside areas, all of which will encourage the designer to make changes to the design. Under these circumstances it is essential that the functional performance as controlled by the critical parameters is not lost. This process provides the necessary disciplines to protect this very important aspect all the time the design is being evolved, whether in the early prototype stages or later during problem solving of field shortfalls. The process is as follows:

1. The value and its possible tolerance as defined by the design (design layout or set of drawings) is calculated and entered against the design intent for every critical parameter in the set. Some of these values and their tolerances will be a straight lift of a dimension and its tolerance from a single drawing. Indeed, these are the most desirable, since they show that parameters that are critical are being closely controlled within a single part. Others will require the calculation of the cumulative effects of a number of dimensions and their cumulative tolerances within the design. The very fact that the process demands the comparison of the drawing with the design intent forces the designer to complete what would normally be called a tolerance analysis, but which is so often forgotten or neglected in the rush to get drawings out to produce the hardware. Further, it answers the inevitable question that is raised by design teams: 'How much tolerance analysis do you want me to carry out?' It is an almost overwhelming task to do a full tolerance check on

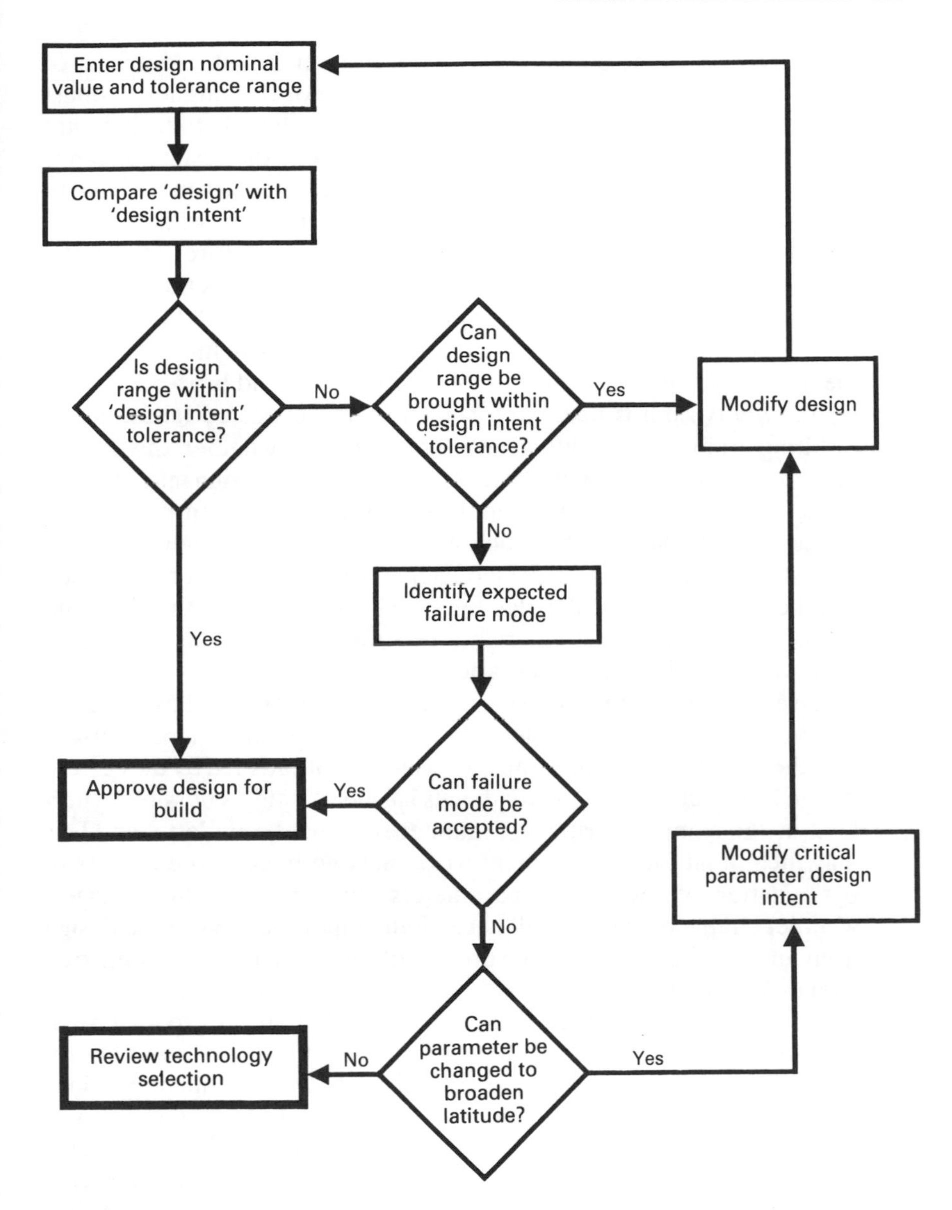

Figure 3.36 Critical parameter implementation process.

anything but the simplest design. Here only those tolerance analyses that affect critical parameters are required to be done.

2. The design is then compared with the design intent. If the range of values shown by the design tolerance build-up is within the range of tolerance shown in the design intent then the design is acceptable. The

amount by which the design sits within the design intent tolerance is of importance. If the design is well within the band, this implies a robust design. In this case it may be opportune to consider whether there are any of the dimensions of the parts that contribute to the design tolerance band that can be relaxed at all, either to make the part more easy to manufacture or cheaper or both. In any event, under these 'acceptable' conditions the design may be approved for the manufacture of the parts. If, on the other hand, the design tolerance band exceeds the tolerance range of the design intent, further questions must be asked. Firstly, 'Can the design tolerance range be brought within the design intent band?' If the answer is 'Yes', modification of the design should proceed accordingly. Otherwise it is necessary to identify the expected failure mode resulting from this condition and then consider whether this failure mode can be tolerated at this time. If it can, then the design may proceed to the build stage only with the understanding that the identified failure will occur and that it must be tolerated temporarily. It is inevitable that it must be corrected at a later stage. Here the chief engineer must decide whether to continue under this known shortfall or take the time and resources now to put matters right. The process allows this decision to be made with all possible facts known.

3. In the event that the failure mode cannot be accepted, one has to look at the only other area for possible change—the critical parameter itself. Can the particular critical parameter which is not satisfied by the current design be modified at all to broaden its latitude? If this is possible, a new look at the complete critical parameter set must be undertaken. This may mean that the development work must go back to the process of optimization of the critical parameters. In any event the changes, whether simple or severe, will necessitate a modification to the design itself and require a re-run of the process of tolerance build-up compared with design intent tolerance.
4. If the review of the critical parameter reveals that no broadening of latitude can occur, this points to a lack of robustness, perhaps a lack of feasibility, of the technology selection. This of course would be a serious result of the process, but it is comforting to note that the methodology of managing critical parameters in this way can at this relatively early stage identify such a serious shortcoming in the design approach. So often a serious situation such as this is not found until *after* the hardware has been built and subjected to testing. The cost of doing it this way is phenomenal compared with a process of identification *before* the hardware is built. Experienced engineers will recognize this situation as not all that unusual and some will look back in envy wishing they had had this process to protect them from such memorable pitfalls!

The management of critical parameters through the design layout and drawing stage does serve to ensure, firstly, that the drawings have been prepared to a standard acceptable to proper functioning of the design and, secondly, that any inherent shortfall in the design itself is likely to be identified before the commitment is made to spend valuable resources on the building of hardware. Finally, any shortfall that is revealed is much more likely to be capable of effective correction due to the wealth of information generated and documented as a routine part of the process.

Critical Parameter Audit Process (Fig. 3.37)

A number of difficulties traditionally arise within the engineering team with the advent of the arrival of the hardware. Firstly, there is the question of inspection of parts. What level of inspection should be implemented, how many parts should be inspected, what percentage of dimensions should be inspected, which dimensions are critical? Secondly, under the pressure of the programme to get into test, there is an understandable impatience to get on with assembly and get the testing under way. This pressure is often so great that hardware goes into test without the team measuring or recording the essential facts about it to identify its initial status.

Managing the critical parameters through the build and prior to test largely eliminates these problems. When hardware is available, either in the form of parts or assemblies, the functional integrity of these parts can be established by confirming that the critical parameters meet the design intent. Note here we did not say 'meet the drawing'. The important thing is that the hardware conforms to the design intent. If it does not , it may either be the manufacturing process that is at fault or the drawing may be incorrect.

1. Each of the critical parameters should be measured on the hardware and recorded individually for that specific piece of hardware. Note that in this case the measurement is absolute. It does not have a range of variance. The values for each critical parameter are then compared with the design intent. If the measurement is within the design intent range, this is accepted, and when all critical parameters are accepted in this way the hardware is approved for testing.
2. Under these circumstances the hardware is now fully audited against the design intent and the parameters are recorded. If later on a failure were to occur, the relevant critical parameters can be rechecked against the values for the start of the test (see critical parameter development, Sec. 3.5). This provides the design engineer with before and after data which can very likely be used to understand and possibly resolve the problem.
3. If the critical parameter does not meet the design intent, the team must decide firstly whether the parts are to the drawing or not. If the parts are not to the drawing, the team must firstly identify what failure modes are likely to be experienced as a result and decide whether these are acceptable at this stage or not. If they are acceptable, a decision to proceed to test may be made with the proviso that the error must be

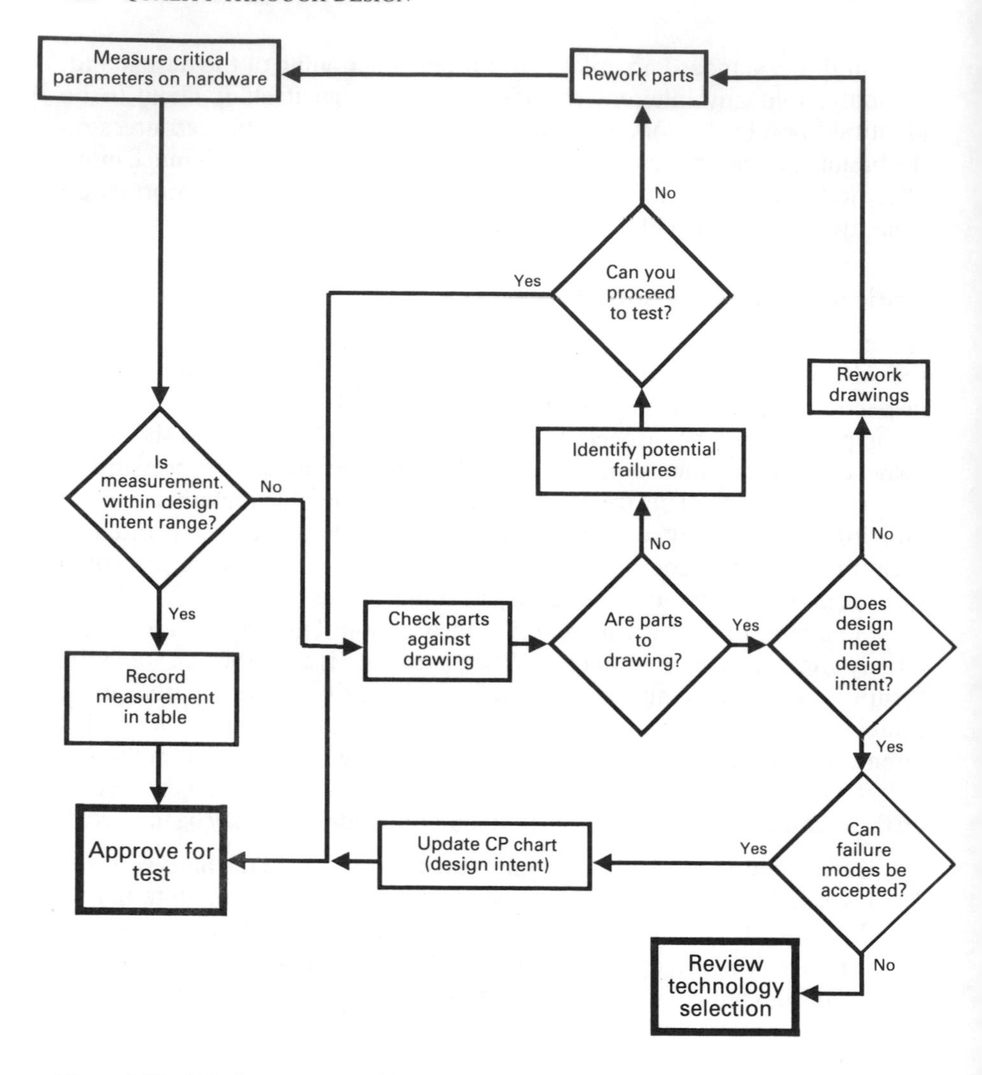

Figure 3.37 Critical parameter audit process.

corrected at a later stage or during the next iteration. If it is not possible to proceed to test the parts must be reworked or remade and then checked through the same process as before.

4. Parts with a critical parameter not meeting the design intent but shown to be correct to the drawing present a more difficult problem, because here is unearthed a fault in the previous part of the process as well as a shortfall in the design process. If the design does not meet the design intent, it is possible that the design and the parts may be corrected. If the design does meet the design intent, a review of the failure modes must be carried out to obtain a full understanding of the function diagram and

its relationships—an exercise that was obviously not carried out properly at some earlier stage. This may result in a modification to the thinking of both the function diagram and the critical parameter set and could very likely highlight a serious problem for the design concept. Most serious of course would be the situation causing a fundamental review of the technology selection for the design.

Once again the critical parameter management process shows itself capable of preserving the integrity of the design through the manufacturing of the parts and the assembly of the hardware. It enables the design team to consistently monitor the quality of the design from a functional point of view and guides the chief engineer in decision making with risk levels clearly quantified. It can also be capable of preventing a design from going ahead with a serious shortfall, one which without this process in place would possibly not be detected until well into the testing stage.

Use of Critical Parameters on Mature Products

Many sceptics will say that there is nothing new in the world of mechanical engineering design. To an extent they are correct. It is rare to come up with a unique concept. Most areas of mechanical engineering design are refinements of something that has gone before. In a way this is the basis of evolution, a safer way of designing than always starting with a clean sheet to develop a more advanced machine.

The need to evolve designs poses problems for the designer when an attempt is made to use the critical parameter management process, because the designer is already faced with a mature design as a starting point which most likely will not have had its critical parameters defined. How then can one start to use the process without going back all the way to redefine the critical parameters for the existing product? The process, I believe, enables the designer to make good use of the existing field data and bring more control of the design to the proposed new product.

Once again the designer starts with the function diagram which depicts function, responses, failure modes and the critical parameters and noise factors that control these. The field data will provide valuable knowledge about which failure modes are prevalent in the mature product and which of these creates the most customer dissatisfaction. This leads the designer into an understanding of which of the functions of the mature machine are least robust or most prone to failure. There is a requirement to link the failure modes with the critical parameters that drive them. A list of these critical parameters provides the basis for deciding where the design effort should be applied to enable the maximum gain to be made in terms of customer satisfaction of the next product.

The designer should review each of the critical parameters so listed and evaluate:

1. How much more latitude can be gained and failure modes eliminated by enhancing the control on the critical parameter.
2. How much of this work has already been done in the time after product launch. Often work is done in the field environment to improve reliability by making post-design modifications. This will have affected the field data that the designer is using in this exercise.
3. How much effort and resource will need to be expended to achieve the gains that result from this study. This must be weighed against the needs of the new product—how much can be afforded, when is it required, etc.

By using this method the designer should obtain a clear set of options showing where the most effective improvements can be made and which improvements, although worthy, will take too much resource or involve too much development work to make it desirable to adopt them.

Design teams charged with the task of improving an existing product often use only their intuition to develop the specific areas to be addressed. This can often cause programmes to get into open ended development which eventually has to be abandoned due to the urgency of getting the new product into the field. Introducing structure to this area by the use of critical parameter management is an effective way of managing the evolution of new product developments from existing designs.

Critical Parameter Status Defined, Pending, Undefined (Fig. 3.38)

One of the difficulties encountered when trying to ensure that the designer can get on with the task of designing has been the confidence in the data that is supplied. It is rare to have the situation where the designer receives information exactly at the rate and in the order that is required for the design. Additionally, it is usually the case that the items that require the longest lead times for the development and manufacture of tooling are the same items that are the most difficult to design early, simply because the input data needed to complete the design is not available. Quite often a designer's only course is to plough ahead with assumed information in order to establish large parts of the integrated design. The danger here is that assumptions can get 'baked in' to the design and become accepted without question only to manifest a functional problem much later on in the development process.

I remember as a development engineer being asked to optimize the thickness of beryllium copper sheet that was to be used to provide a flexible surface to a copying machine photoconductor. Everything being used at the time was 0.004 in. thick and my first task was to find out why this thickness had been chosen in the first place. Detailed investigation of the history of this particular component revealed that the first photoconductor had been made by a technician who happened to have raw material in stock that was

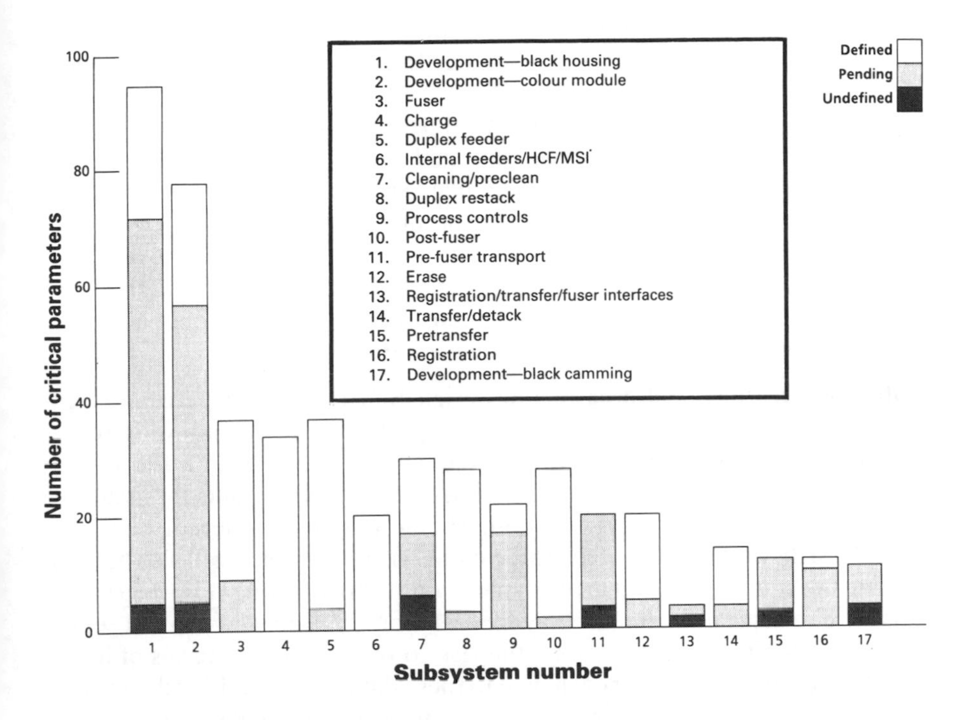

Figure 3.38 Critical parameter status summary.

0.004 in. thick. Having been tried once successfully, this material was assumed to be optimum and nobody had ever questioned it again!

As we have seen, the use of critical parameters as a vehicle to define the design itself can serve to resolve this problem and help to describe the confidence level associated with particular areas of data.

The list of critical parameters, their nominals and their tolerances serve to represent the design intent. This set of data for any functional area is continually being updated as optimization occurs and further refinements are made due to the experimental data being generated. If the design process were to wait until all these data were completed and stable before the design was allowed to start, the process would become serial and its duration would be so long as to be unacceptable to the programme and uncompetitive. The design must therefore be started without the full information available, but must not be finalized for manufacturing until the complete set of data is known.

In order to define clear confidence levels in the data, each of the critical parameters is assessed as to its level of stability. This stability is defined by one of three categories:

1. Defined. This means that the critical parameter is stable and has been

optimized either during the current programme's experimentation phase or following mature knowledge from its similar use in a previous programme.

2. Pending. This means that the critical parameter is in the midst of being optimized and that although nominal and tolerance values for this particular critical parameter have been stated, these may be subject to change as the optimization work continues.
3. Undefined. This means that the critical parameter has been evaluated only at the minimum level, that it is very likely to change and that any figures stated for its nominal value or tolerance are for reference only and not based on any engineering evaluation.

Obviously a functional area of the design that has a high percentage of critical parameters in the 'defined' category can go ahead with some confidence. On the other hand, a functional area with even a few in the 'undefined' category must be treated as very immature and the issue of drawings for manufacturing or tooling should not be considered.

A further refinement of the three categorizations which will also help to define the confidence level in the data is to add a 'risk' rating (Fig. 3.39). For instance, a critical parameter may be 'pending', but may have a full set of data identified with it. By stating the risk to the function in terms of high, medium or low in the event of change to these data gives an added dimension to the assessment of confidence. Thus a designer looking at data described as 'pending, low risk' might be inclined to go ahead with this immature data on the basis that a subsequent change may not incur a major impact on the function of the design. Another additional feature to this information is to add the dates when the team expects that the critical parameters will become 'defined'. Depending on receipt of this information, this enables anyone to compare the expected date with their own plan and make any compensations necessary.

The diagrams show examples of typical data that might be supplied to a designer. Clearly this gives the designer a much better understanding of what can and cannot be committed at any stage. Furthermore, the chief engineer, when assessing the status of the overall programme, can use this type of data, probably in condensed or summary form, to make key decisions on the programme. The engineer can choose to take risks, but when this is done it will be based on good information from the working level of his team.

3.6 VALUE ENGINEERING

Value analysis and value engineering are not the same as cost reduction. Value engineering attempts to enhance the value of a part or assembly by getting more function out of it for the same, or less, cost. Cost reduction

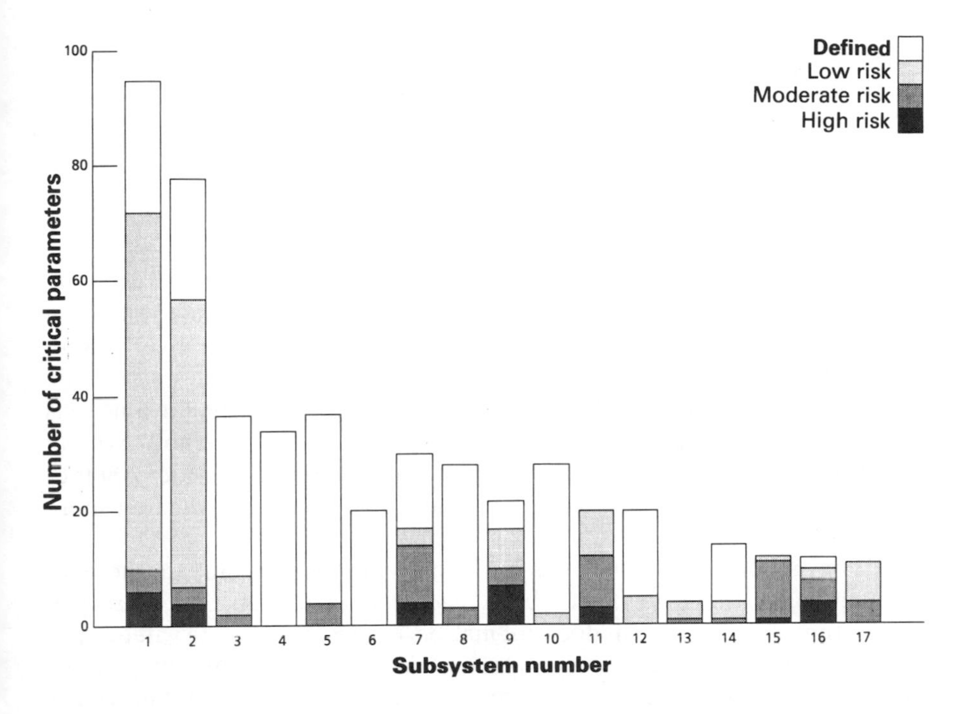

Figure 3.39 Critical parameter status summary showing risk.

does something different. It merely tries to reduce the cost of the part or assembly without reference to its function.

To illustrate the difference between cost and value, let us use a simple example. If someone were to offer you a glass of water, how much would you be prepared to pay for it? In normal circumstances, probably not very much. Since you are not particularly desperate for a drink you would not want to part with good money for it and anyway you would want to choose your preference for a drink if you were to pay. The value and cost of water in these circumstances is low in both cases. Now imagine that you are offered a glass of water for sale in the middle of a desert where you have been marooned for some hours, and where your life is in danger for lack of water. Now the value of a glass of water is enormous. Its cost is still low, since in both cases it fell from the sky in the form of rain and had only to be collected. Therefore cost relates to how much effort is required to obtain a particular commodity while value is dependent on the circumstances in which it is used.

When we talk about value engineering we are concerned with how it is used as well as how much it costs. More importantly, we are concerned with the relationship between these two things.

FAST diagrams can be a valuable tool to study these things and apply value engineering principles. Value engineering is normally applied once a design has been established, but the same principles can also be applied early on in the design process with great success.

Value engineering is addressed through four phases, namely: definition, innovation, selection and implementation.

Definition

Let us refer to the FAST diagram previously developed for the desk-top stapler (Fig. 3.20). The diagram forms a picture of the linkages between each of the functions of the stapler, each being described in a verb–noun form. Just the activity of drawing the diagram enables us to make some attempts at optimization of the system. We can look for similar functions and see if any of these can be combined in the same part or assembly. We can understand relationships between any two functions and consider whether either of these could be changed to make the combined function more effective.

For any one of the functions in the diagram we can estimate the cost. Depending on the stage that has been reached in the design, the accuracy of our estimate will vary. This does not matter. In fact the absolute value of our cost estimate in each case is not important. What is important is the relative costs between one function and another. For this reason it is not necessary for the design to be well defined and this is why the process can be effectively applied, even in the early stages of design.

Let us take the simple function 'open throat'. Our FAST diagram shows us that this is achieved by providing a pivot, a spring (to provide torque) and a stop. By extending the FAST diagram further to the left we can begin to convert these functions into the set of parts that must be provided to support them. Each of these parts in the assembly can have a cost estimate placed against it. The parts are:

Hinge bracket
Lift spring
Stop

Let us assume that the costs associated with these parts are the full costs that go to provide the desired function 'open throat'. The costs are:

Hinge bracket	4p
Lift spring	5p
Stop	1p

Therefore the total cost of providing the function of opening the throat is 10p. Now we can ask ourselves whether that cost provides reasonable value for delivering that function. The following steps are taken:

1. The value engineering process demands that a full understanding of the functional relationships within the system is in place. This can be accom-

plished by drawing a FAST diagram, and by extending the FAST diagram to the left the individual parts can be defined. The diagram can be used later to review the relationships between functions and in particular to assess whether the parts that are defined to enable the system to function are the best ones for the job or whether they are configured in the right way to deliver the function in an optimum way.
2. Confirm using the FAST diagram that the parts that one has in the physical design are all accounted for in the FAST diagram. If they are not, then study the FAST diagram to understand how the design differs from the functional relationships depicted by the FAST diagram. Similarly, if there are more parts listed by the diagram than are in the design, then this must also be rationalized. It is important that the FAST diagram and the design are compatible and represent each other, the one physically and the other conceptually.
3. To each of the individual parts add the cost. Write each of the costs in the box representing the part on the FAST diagram.
4. Then move to the right along the diagram, adding the parts costs to give the costs of each of the functions. As the process proceeds to the right of the diagram the functional costs are summed cumulatively, and eventually the final cost entered into the right-hand box represents the cost of delivering the function of the system.
5. Now one can begin to apply the judgement of value to each of the boxes in the diagram. Consider each of the functions and ask whether the cost in each case is reasonable considering the value of the function. Then compare functions and assess whether the comparative costs are reasonable in each case. Select functions that should be the subject of a cost reduction exercise. Make these targets for value engineering.

Innovation

The search for new ideas or solutions to problems can take many forms. None of these is better than another and it is up to the reader to select the best method of generating ideas, either to suit the individual type of problem or to suit the individual problem solver or team.

There are a number of techniques described elsewhere in this book. The list is as follows:

1. *Brainstorming*. This is good for getting a long list of ideas which can come from obscure areas of the mind and which are therefore likely to be most innovative. The process generally develops a very long list of ideas, many of which have to be eliminated. Quite often, however, a good solution emerges from this technique.
2. *Problem solving cycle*. This proves to be excellent when a specific problem can be defined with clear boundaries and lots of data to

quantify it. A detailed description of the process can be found in Sec. 3.8.

Selection

The selection of options following value analysis is a most important part of value engineering. It should be appreciated that no selection process will be foolproof. More importantly, the process should be capable of revealing the good and bad points of the various options in a way that will provide the team with a deeper understanding of what is at stake. Referring back to Sec. 2.2, both the Pugh process and the Combinex method are excellent processes for making selections. The choice between these two processes is entirely dependent on the level of detail of the study. Sometimes a significant amount of numerical data is available to quantify each of the options. At other times the various ideas, although all at equal levels, may have no numerical data at all associated with them. The Pugh process should always be used when there is little quantitative data available or when different ideas are at different levels of evaluation. Whenever there is a significant quantity of information in numerical form, the Combinex method should be used.

Implementation

The implementation phase of value engineering makes the previous phases look easy. Implementation in general is difficult because it generally means change, and individuals resist change by nature. In addition to this natural reluctance to implement change there is the knowledge that the implementation of a new idea is going to involve risk. Risk means the possibility of individual failure and possible embarrassment. The person responsible for the implementation of the idea must therefore be sufficiently motivated by the opportunity that the new idea offers to offset the reluctance to bringing in change and risk. Hence implementation is most likely to occur if the magnitude of the change and the risk can be minimized and if the opportunity can be maximized.

Minimizing the magnitude of the change Firstly, the change must be fully quantified and evaluated. A change that is vague is quite likely to appear complex, whereas a change that has undergone a full analysis and has been evaluated together with its interfacing areas will have more credibility, even if it appears complex. The designer who will implement the change will want to balance all the requirements equally so that no area suffers more than another. Therefore presentation of the change in a way that takes into account all the impacts will be more palatable.

Similarly, ill defined changes may look original but will lack that robustness which will give the designer confidence to incorporate the new idea.

Minimizing the risk It is essential to convince the designer who will implement the change that the risk is far outweighed by the opportunities. Again a full analysis of the benefits of the change, ideally in numerical terms, will provide this confidence.

Maximizing the opportunity It is a good idea to list the benefits of the proposed idea and to show how they solve the problems of the existing design. Obviously, there will also be disadvantages and these must be listed. However, if the correct selection process has been followed, the disadvantages should not be prohibitive. If they are, then this may be a sign that the value engineering process in this case has not yielded the best solution. There may be some peripheral effects of implementing the solution that are advantageous, such as improved lead time on the tooling or improved maintainability. These should also be listed to emphasize the opportunity.

The implementation phase of the value engineering exercise is very important because without an implemented solution the rest of the value engineering process will have been wasted. To promote implementation the designer who will effect implementation should be involved from the very beginning of the task. The elements of change, risk and opportunity should be carefully considered as they relate to the designer.

An Example of Value Engineering

The stapler shown in Fig. 3.40 is an excellent example of value engineering. The device has embodied all the necessary functions into a much smaller number of parts. The stapler actually uses only eight parts to deliver the same functions as the desk-top stapler described previously, which had 27 parts (Fig. 3.41).

Referring to the FAST diagram (Fig. 3.42), it is easily seen how the design of the stapler has been efficient in the use of parts shown shaded to deliver maximum functionality. Let us look at some of the value engineering aspects that have been employed:

1. *Anvil pressed into base.* Instead of treating the anvil as a separate part it has been combined into the pressing of the base.
2. *Single pivot for all vertical movement.* The vertical movement required of both the striker plate and the opening of the throat has been enabled using a single pivot. The top, which drives the striker plate, and the staple support, which moves up to create the open throat, have both been pivoted by the same pin, which holds four parts together using concentric holes.
3. *Striker plate and leaf spring combined.* The leaf spring, which is used to open the throat, also forms the striker plate itself. As a result, the spring action brings the striker plate up to its rest position after stapling. The spring force for opening the throat has been created by allowing two

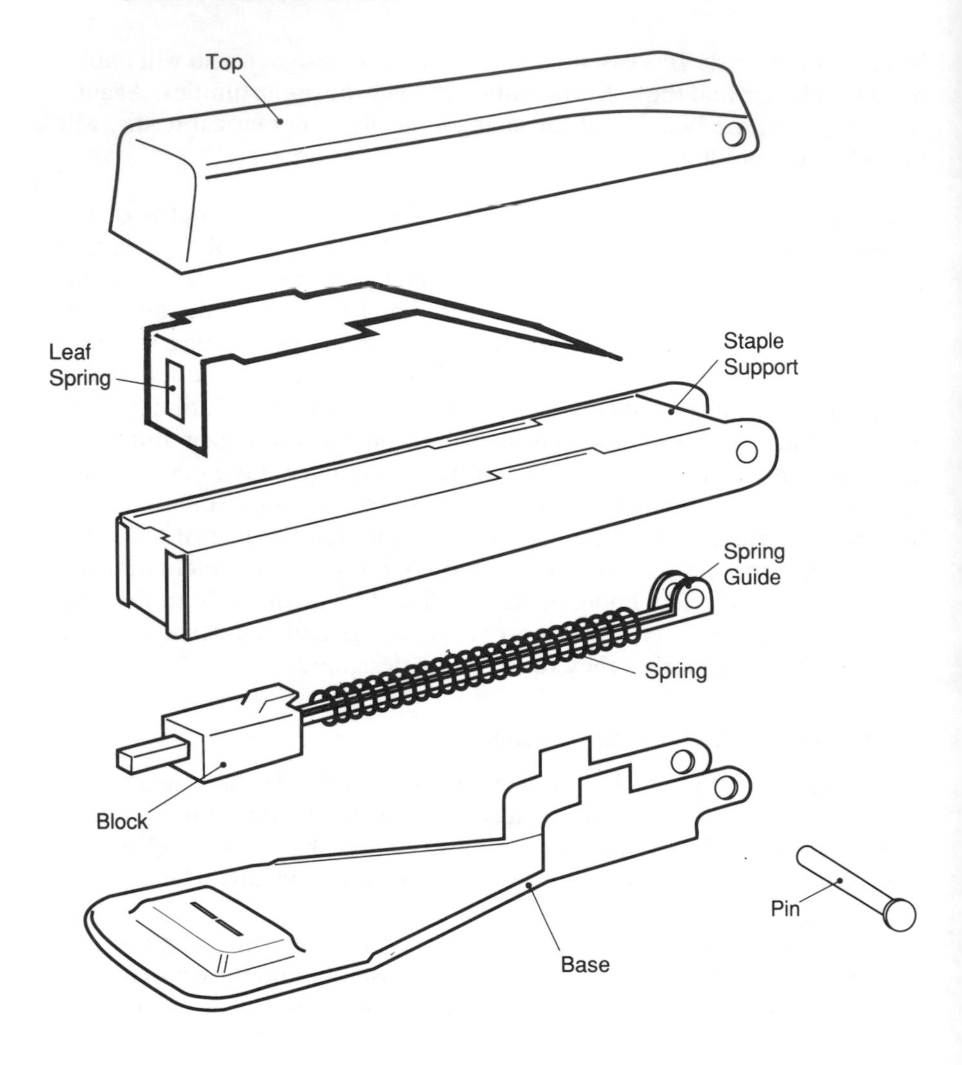

Figure 3.40 The value engineered stapler.

protrusions from the base to pass through slots in the staple support and engage with the leaf spring.

4. *Stop for striker plate embodied in staple support.* By turning out a tang in the pressing of the staple support and using a slot in the leaf spring, a stop has been created for the upward movement of the striker plate.
5. *Striker plate guides combined in staple support.* The guides for the striker plate have been combined into the pressing of the staple support. It is interesting to note that the striker plate does not need to be constrained in the vertical plane. This constraint is already automatically achieved by using the leaf spring as the striker plate, which by its

Number	Part	Cost (£)
1	Top	0.28
2	Base	0.35
3	Staple support	0.32
4	Leaf spring	0.26
5	Block	0.05
6	Spring guide	0.20
7	Spring	0.08
8	Pin	0.03
	Assembly	0.15
	Total	1.72

Figure 3.41 Parts and cost of the value engineered stapler.

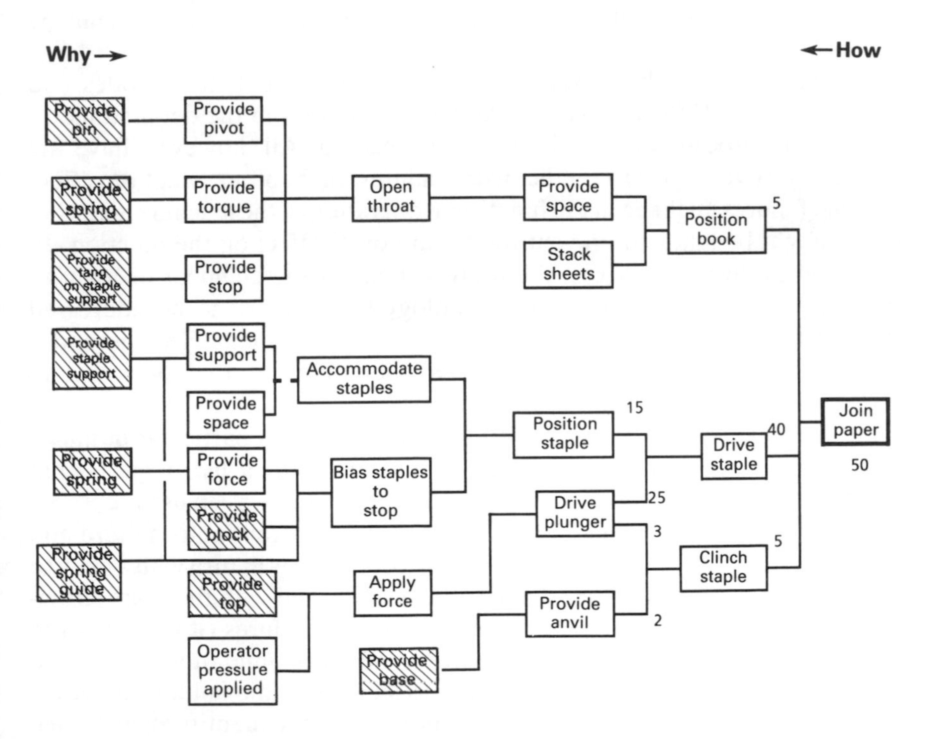

Figure 3.42 FAST diagram for the value engineered stapler.

very nature gives rigidity to the vertical movement. The difficulty of this alignment is often a major cause of unreliability in a stapler. Thus the value engineering in this case has not only provided a simpler and cheaper design but also one with greater reliability.

The use of FAST as an aid to understanding function and value engineering is invaluable when designs are being developed. The example of the simplification of a stapler serves to illustrate the principles of this. In practice the process will pay handsomely in terms of improving reliability and reducing cost.

3.7 FAILURE MODES AND EFFECTS ANALYSIS

The use of the critical parameter management system has enabled us to see the exact meaning of failure in terms of a function. We have seen that a failure is not necessarily a catastrophic event, as the more popular definition of failure would imply. Rather, it is a consequence of not meeting a customer's requirements. When we look at the function diagram it is obvious that the level of the response at which we define a failure is entirely dependent on our understanding of the customer's requirements. Any change in customer requirements will inevitably lead to a redefinition of what we mean by failure.

Most engineers have heard of the process called failure modes and effects analysis (FMEA). There are many variations in the detailed methodology of processes that fall under this heading. All, however, have the same objective, that is to predict what failures might occur, what the effect of such failures might be in the functional operation of the machine and what steps might be taken to prevent the failure or its effect on the function. In addition to these objectives, the method enables the engineer or team to identify areas of the design or technology that require to be addressed further to enable reliability advancement of the technology. The method is particularly useful in this respect when applied to immature designs that do not yet have proven technology readiness.

The method is based on the list of functions that a particular design is intended to perform. Against each of the functions a list of parts that contribute to that function is generated. (Here, again, we can use the FAST diagram.) Each one of these parts will have some potential for failure and the next stage in the process is to study each of the parts in turn with a view to listing every potential failure that could be associated with that part. Obviously some parts will have a long list of potential failures (if the list is too long, perhaps we should look again at the design!) while others will have very few or zero. The important thing is to identify all the failures, however trivial or repetitive this may seem. The next stage is to identify against each of the failure modes the effects of that failure on the functional operation of the design and then against each of those the root causes of the failure. One important aspect of this process at this stage is that the designer can now ask what controls have been put on the part under investigation to ensure that the failure has the lowest probability of occurring. For instance, in the case of a rubber coated roller, one potential failure mode may be the debonding of the rubber from the metal core as this would cause a malfunction in the machine. The designer can ask, 'How can I ensure in the design that the potential for the debonding of the rubber from the core is low?' The designer may decide to specify a coating procedure on the drawing which shows the assembly of the rubber to the metal core. In this way the designer will be able to ensure that good quality control of the production process is in place. Additionally the designer may decide to call for a test on the effectiveness of the bond once it has been accomplished either on all rollers (non-destructive) or on regular roller samples (maybe a destructive test). At this stage in the process the analysis of failure modes, their effects and the parts that are involved provide a kind of check-list for the designer to enable the closest possible controls on the design to be applied.

When considering failures and their effect on the function it is important not to give each the same level of importance. What one really wants to know is what is the order of priority in which these failure modes and their effects should be addressed. Again this is very much dependent on the impact that the failures are likely to have on the customer. In order to evaluate this, three factors are taken into account (Fig. 3.43):

1. The severity of the failure in terms of customer dissatisfaction. A customer may not react immediately to a failure that perhaps causes the device to become more noisy, but will obviously react strongly to a failure that puts him in danger. These are just two extreme examples of how the failure and its effect must be measured in terms of how the customer perceives it.
2. The probability of the occurrence of the failure. This is dependent on the type of failure and the robustness of the design in which it occurs. It will also be dependent on the product's age and its position on the traditional reliability/time curve, often called the bath-tub curve.
3. The probability of the design or development process detecting the failure. Some failures may not be easily detected by the development process. A failure internally whose effect does not manifest itself in the output performance of the machine may not be readily identified by the testing programme due to the need for perhaps longer tests than can be accommodated or the inability to observe the failure.

These three categories, severity, occurrence and detection, can each be rated (here as an example out of 8) to assess their level using the judgement of the engineer or perhaps by using information from products already in the field or on test. The product of the three numbers is called the risk priority

<table>
<tr><th>Rating</th><th>Severity</th><th>Occurrence</th><th>Detection</th></tr>
<tr><td>1</td><td>Exceeds specification but not noticed by customer</td><td>Never</td><td>Very high—programme design process will detect failure</td></tr>
<tr><td>2</td><td>Noticed by customer but does not affect the product function</td><td rowspan="2">Very occasionally</td><td rowspan="2">High—programme design process is likely to detect failure</td></tr>
<tr><td>3</td><td>Noticed by customer, minor effect on product function, customer accepts condition</td></tr>
<tr><td>4</td><td>Customer dissatisfied with function of product</td><td rowspan="2">Occasionally</td><td rowspan="2">Medium—programme design process may detect failure</td></tr>
<tr><td>5</td><td>Significant effect on customer satisfaction</td></tr>
<tr><td>6</td><td>Significant inconvenience to customer</td><td rowspan="2">Frequently</td><td rowspan="2">Low—programme design process is unlikely to detect failure</td></tr>
<tr><td>7</td><td>Significant annoyance to customer</td></tr>
<tr><td>8</td><td>Customer endangered</td><td>Very frequently</td><td>Zero—programme design process will not detect failure</td></tr>
</table>

Figure 3.43 FMEA rating chart.

number (RPN), and the value of this number can be used to help prioritize any activities initiated to improve overall reliability. For instance, a low RPN may be regarded as a problem of low priority whereas those of higher RPN may be regarded as problems having a major impact on customer satisfaction.

It is very important to develop a specific set of criteria for the development of the RPN in FMEA for a particular industry. The absolute levels are not so important as long as the relative levels represent the priorities and preferences of the team. One essential task is to develop charts that represent these criteria for your own business. An example is given here, but although this can be used as a guide it is better to develop unique tables of your own. Incidentally, it is not essential to rate the S, O and D levels out of 10. In the example shown a maximum of 8 was found to be more appropriate. This of course gives an RPN range from 1 to 512. Any score of 8 in the severity column should alert the team to a dangerous problem, perhaps threatening the customer's safety, and it may be prudent to mandate that such a score requires immediate attention and maximum resources in solving the problem.

Identifying any fixes that can be applied to the failure areas and comparing RPN ratings helps the designer to see where there are significant reliability problems with little or no fix effectiveness available. Often this will lead the thinking into identifying areas where new investment and a search for new technologies must be applied.

The Steps in the FMEA Process

The steps in the FMEA process are similar whether the FMEA is for a systems level design, a design defined to parts level or a process FMEA during the manufacturing cycle. The steps defined here relate to a 'design FMEA', that is working bottoms-up from the list of parts in the design. The sequence of steps in the process enables the table of Fig. 3.44 to be filled in order, from column 1 to column 11.

Step 1 Enter the part name and number. Each part must be considered individually and assessed for its potential to fail.

Step 2 Enter the function of the part and what it is required to do. Some parts have more than one function. In this case it is necessary to break down the individual functions separately and consider the failure mode against all the defined functions. The best way to describe functions is to use a verb–noun combination, just as was done when developing FAST diagrams. FAST diagrams could be employed usefully here to help to identify the functional relationship with the part and where that fits in the overall system so that effects from the failure may be more easily determined.

Step 3 For each of the functions listed in step 2 enter the potential failure modes. This will further expand the table because there are likely to be many potential failure modes associated with each of the functions.

Step 4 For each of the failure modes identified in step 3 describe the consequences or effects of the failure. Here look for effects that may be manifested in the assembly, the overall system, those that might impact the customer and of course any that impinge upon any regulations such as safety or environmental requirements. Try to stretch this thinking as far as possible by thinking of what happens next as a result of the failure under consideration.

Part	Function	Potential failure mode	Potential effects of failure	Severity	Potential causes of failure	Occurrence	How will the potential failure be detected?	Detection	RPN	Actions
Outer tube	Provide grip for writer	Hole gets blocked	Vacuum on ink supply stops flow	7	Debris ingress into hole	3	Check clearance of hole	5	105	• Make hole larger • Remove cap
Ink	Provide writing medium	Incorrect viscosity	High flow	4	Too much solvent	2	QC on ink supply	4	32	• Introduce more rigid QC
Ink	Provide writing medium	Incorrect viscosity	Low flow	4	Too little solvent	2	QC on ink supply	3	24	• No action required
Ball and seat	Meter ink supply	Incorrect fit	Ball detached	8	Total failure	2	Inspection checks	2	32	
Ball and seat	Meter ink supply	Incorrect fit	Ball loose	6	Blotchy writing	3	Sampling checks	6	108	• Introduce in-process checks during manufacture
Ball and seat	Meter ink supply	Incorrect fit	Ball tight	7	Intermittent writing	4	Sampling checks	6	168	• Introduce in-process checks during manufacture • Control ball and seat variation
Inner tube	Contain ink	Tube kinked	Ink flow restricted	5	Poor handling in manufacturing	2	No current checks or tests	8	80	• Introduce detection checks for this failure
Plug	Close outer tube end	Wrong size	Cannot be fitted	2	Moulding process not in control	2	During assembly	1	4	• No action required
Plug	Close outer tube end	Wrong size	Falls out	4	Moulding process not in control	2	No current checks or tests	8	64	• Eliminate part • Control part moulding process variations

Figure 3.44 Failure mode and effect analysis table.

Step 5 This step uses the customized severity table to make an assessment of the numerical rating which can be attributed to the most serious effect due to the potential failure. This numerical value is now entered in the table.

Step 6 Enter the potential causes of the failure mode. Here one should try to emphasize the root cause of the failure. This can be done by asking why, why, why, . . . sequentially until a root cause of the failure is found. As an illustration of this, let us consider an optical sensor operating within a system to detect the presence of light or dark. (This is often used as a microswitch, the light/dark state being achieved by a flag across the sensor driven by an actuator, similar to that of a microswitch.) Let us assume that the potential failure mode defined is that the sensor goes 'open circuit'. The potential effects of this failure might then be an erroneous signal to the control circuit. The effects of this on the system can be expanded to identify the effect on the customer. The potential cause of the open circuit must now be identified. Perhaps it is due to a fractured wire. Now ask 'why'. The fractured wire may be caused by too much flexing of the cable. 'Why?' Perhaps there is not enough support to the cable in the areas of movement. Again 'Why?' Is there limited space? Are the supports inadequate for the size of wire? By continuing this line of thought the engineer can get a much better understanding of the root cause of the potential failure.

Step 7 Enter the occurrence rating for each cause of the failure that has been identified in step 6 using the customized occurrence table.

Step 8 Now the engineer has to consider how the process which is in place can help to detect the potential failure mode. In the column list the design process evaluation procedure which will be used to detect the cause of failure listed in step 6. This might be a particular test or check. It might rely on supplier technical information about a particular component or it might identify design calculations that could be used to predict failure.

Step 9 Enter a rating from the detection part of the table which reflects most closely the ability to detect the cause of the failure which has been identified in step 8.

Step 10 Calculate the risk priority number (RPN). This number is calculated by taking the product of the severity, occurrence and detection ratings, or

$$\text{RPN} = \text{S} \times \text{O} \times \text{D}.$$

Step 11 Assess the most important FMEAs by comparing the RPN scores. Decide which set of these should be actioned first. This will depend on the resources available and the seriousness of the overall FMEA picture. Define actions against each of the FMEAs, assign ownership to each problem and start to prepare action plans to address the activities. If there is already a problem management system in progress, tie the main actions into this system, thereby ensuring the management of only one system.

Above all, make sure that the documentation for this exercise is in place and can be readily employed for future products.

FMEA can be applied both to the design of products and systems and also to processes. In the area of product development and manufacture it is likely that there will be two FMEA exercises for the same area of the product, one initiated by the design team and one initiated by the production team. Ideally these studies should both be carried out by multi-disciplined teams comprising members from both the design area and manufacturing. This encourages the teamwork for the product, encourages effective communication across disciplines and ensures joint ownership of the outcomes.

Let us look at an example of a typical design FMEA applied to the ball-point pen for which we developed a FAST diagram (see Sec. 3.4). The FAST diagram (Fig. 3.23) represents the functional operation of the pen and in the boxes shown shaded depicts each of the components that go to make the assembly of the pen. Drawing this diagram is the first step in applying FMEA. In a more complex system, the FAST diagram would be larger or composed of a series of separate FAST diagrams, and in these cases FMEA may need to be applied to a more limited area in order to make it able to be handled.

The table for studying and recording FMEA shown in Fig. 3.44 lists:

Each part
Their potential failure modes
The potential effects of the failure
The severity that the effect of that failure would have on performance
The potential causes of the failure
The probability of that cause occurring
Any controls that can be put in place to minimize the occurrence or severity
The likelihood of the failure being detected during the design / development process
The RPN product (S × O × D)
Any actions that should be taken to minimize the risk of failure

Let us look in detail at the first line of the table.

The part we are concerned with here is the outer tube of the pen, which is the part that the writer holds and which supports the inner tube and the writing head. One of the requirements of the outer tube is that it should not be totally sealed against the ink in the inner tube, otherwise a partial vacuum may cause ink flow to be restricted. The design therefore calls for a small hole to be incorporated in the side of the tube to minimize this condition.

When considering potential failure modes it is obvious that one of these may occur if the small hole becomes blocked. This may happen during manufacturing or while in the care of the customer. Whenever it occurs the effect of the failure will be the same for the customer, that is the partial vacuum may restrict flow of the ink and degrade the writing performance. If the failure occurs, how severe will be the effect on the customer? On a scale of 0–8 it is estimated that the severity will be 7, that is quite severe although not catastrophic.

What are the potential causes of this failure and how likely are they to occur? Here it is estimated that some debris could find its way into the hole to block it, but that this is not very likely to happen; again on a scale of 0–8, 3 is estimated.

Are there any controls that could be applied to test for the occurrence of the failure or to limit the possible occurrence of the failure? Perhaps high pressure applied during the assembly process would ensure that any debris in the hole at that stage would be blown out and cleared otherwise visual inspection prior to shipping is the only other way. With the checks currently in place in the process, how likely is it that the failure will be detected? This is estimated at a level of 5, meaning that the test processes in place may detect the restricted flow of ink. This would be dependent perhaps on the usage rate of the pen during checking, intermittent usage allowing the partial vacuum to recover.

Multiplying the three ratings together gives an RPN rating of 105. Continuing through the exercise for each part in turn gives a series of RPNs which can now be compared to establish priorities.

If we look at the results of the FMEA we see that the highest RPN rating occurs in the area of the ball and its seat, namely a rating of 108 and 168. The question now is what can we do in design to minimize this risk? It is possible that closer tolerancing of the mating parts in this region may help this situation, but this may also put up the cost. In manufacturing there could be more in-process checks introduced to establish adequate fit of the two parts.

Looking at the next highest RPN, the potential failure due to the blockage of the hole in the outer tube could be minimized by making the hole larger. Should the design eliminate the plug and dispense with the hole? It would eliminate the problem and save parts and cost!

This example is a very simple one, but it serves to show how the attention of designer (and manufacturing engineer) can be focused towards the areas of highest risk in a design with the potential to apply effective solutions.

3.8 PROBLEM SOLVING

The process of problem solving is often taken very much for granted by people who are in the business of finding solutions as a part of their day-to-day work. Problems come in many forms and are solved by many different remedies. However, they do not solve themselves just by having attention given to them. They are solved by people and they are solved by people most successfully when they are addressed in a systematic way.

Problem solving is hampered by a number of difficulties, all of which can be eliminated if the problem-solving activity is regarded as a process and when such a process is followed in a clear and thorough way. To give some ideas as to what can hamper the effective solving of problems, here are some of the pitfalls that may be encountered by the untrained or unsuspecting problem solver:

1. *Jumping to a solution.* A discussion aimed at solving the problem can often be seen to jump to a solution before the exact definition of the problem has been described. Even if this finds one solution it will almost inevitably narrow the range of solutions available.
2. *Lack of gathering critical data.* Taking shortcuts and not obtaining the full data that are available and therefore depriving the problem solver of a thorough analysis of the problem will inevitably reduce the capability to come up with the best solution to the problem.
3. *Lack of control over the basic aspects of the problem.* Sometimes a would-be problem solver or team will be diverted into working on problems that are outside their range of control. Solutions in these cases are therefore outside the grasp of the team or unimplementable as a result.
4. *Working on problems that are either too general or too large.* Here the size of the problem may be too extensive or broad for the problem solver to cope with. It would be better to break down the problem into manageable pieces. The same applies if the basic problem is found to be too much of a generalization.
5. *Proposing a solution that has inadequate justification.* If a solution is found, it can only be a real solution if it can be justified on the grounds of cost, timeliness, etc. If such justification cannot be provided, then the solution is as useless as if it had not been found.
6. *Not involving the people who will be involved in the implementation of the solution.* If the people who must be involved in implementing the solution do not support it then the implementation may not be possible. Their agreement must therefore be a key factor in the solution itself.
7. *Lack of planning for the implementation of a solution.* Finding a solution without also developing the specific plans for how it will be implemented and evaluated can prevent effective execution of a solution.

The Problem Solving Cycle

The problem solving process proposed here is not complex, but if followed it will effectively eliminate most of the pitfalls quoted above. It can be applied to all problems whether technical or otherwise and can be used with equal effectiveness by either individuals or teams. The process identifies six steps to the problem solving process (Fig. 3.45). These are:

1. Identify the problem
2. Analyse the problem
3. Generate potential solutions
4. Select and plan a solution
5. Implement the solution
6. Evaluate the solution

Identify the problem Step 1 of this process is one of the most important because if it is done inadequately the remaining five steps of the process are likely to be less effective or at the worst deliver incorrect solutions. Although this is one of the most important steps it is one that is not always given sufficient time. It should be stressed here that the time spent on this part will pay handsomely in the later stages of the process. Insufficient time spent on step 1 may seriously affect the useful progress of the rest of the process. The definition of the problem provides the focus of the true causes of the problem and enables the subsequent development of the solutions to have a firm foundation.

It is essential during this step that the problem solvers are able to distinguish between the problem itself and the symptoms of the problem. If a fuse blows in an electrical appliance this indicates that there is a problem. However, the blowing of the fuse is not the problem. Whatever caused the fuse to blow is the problem and the fuse failure is just the symptom. Problems that are too broad, too complex or not within the control of the group should be avoided and broken down into more manageable problems. Similarly, problems that are too difficult for the group to solve without additional resources or those that will take too much time should be avoided.

Analyse the problem Having the discipline to analyse the problem once it has been defined is almost alien to human behaviour. It seems more natural to go in search of a solution once one knows what the problem is. However, looking at it logically it is more sensible to really understand what is behind the problem before jumping into a solution. This phase is aimed at doing just that. It is necessary, however, to warn that even when one is aware of the need to analyse the problem there is a great temptation to search immediately for a solution and one should constantly be on one's guard to identify when this is happening.

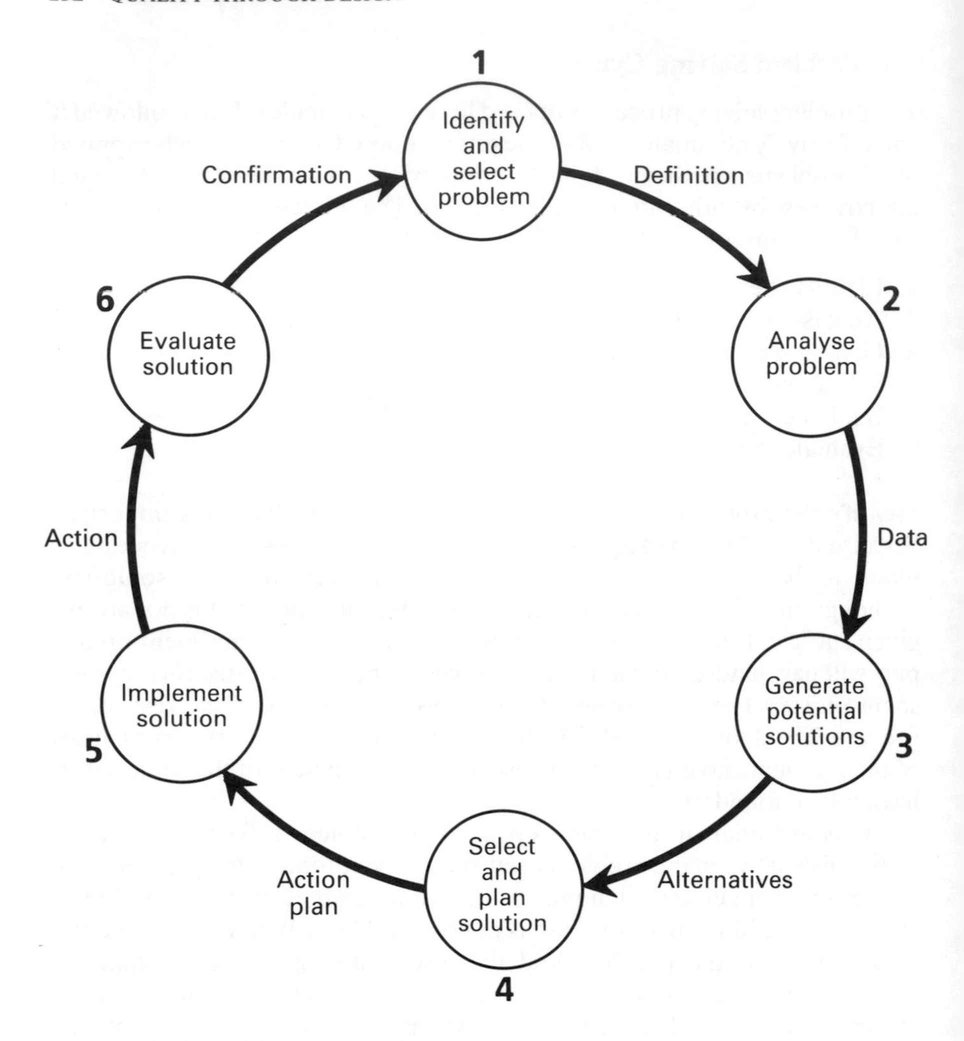

Figure 3.45 The problem solving cycle.

The process of analysing the problem means the collection and display of data so that the problem solver or team can really understand the elements of the problem and how they create the discrepancy that causes the problem. Here is a list of methods and devices for displaying and quantifying data connected with the problem. Obviously not all will be used for one problem, and their use must be tailored to addressing the aspects of the particular problem under investigation. The output from step 2 is data. This can be compiled using a number of different techniques.

Checksheets Checksheets are used to determine how often an event takes place over a given period of time. Their purpose is to track, not analyse, data and can often help to highlight what the essence of the problem is. They can be used to record:

The number of times an event occurs
The length of time for an event to happen
The frequency of occurrences
The effect of an action over a given period of time

Other devices help us to display in a graphical form the data that have been collected.

Graphs These simply convert the data into a graphical form and help the team to see important discontinuities that may be related to the problem.

Pareto charts A Pareto analysis is a bar chart which has the bars rank ordered in importance. In other words, it orders the information from the largest to the smallest. This serves to draw attention to the most important element (or elements) of the problem, and can often direct the team into concentrating on those parts that will give the greatest return for effort, rather than diluting the effort by attempting all elements, or the wrong ones.

Flow diagrams Flow diagrams enable the display of a sequence of events and make it easier to look at processes and procedures, especially those that are complex. They may highlight errors or discrepancies in the process.

Fishbone diagrams These diagrams enable the user to piece together the causes that together go to make up the effect. They enable the user to group causes together and encourage the search for causes that may not be immediately obvious.

Family trees These diagrams can be used to link as a hierarchy the elements that combine to deliver an error or failure. Once drawn the user can look for repetition of elements which would indicate a significant cause of the problem.

Force field diagram This diagram defines the gap between what is and what should be. It identifies all the forces that are helping to close the gap and the forces that are helping to maintain or extend the gap. By considering reducing or eliminating those forces that tend to extend the gap and considering increasing those forces that tend to close the gap, a greater understanding is gained as to why the gap exists and how it might be reduced.

These are just some of the ways in which, with the data available to us, we can display it in such a way as to help our understanding of the problem or get some guidance as to where we might seek solutions. More details of these methods are given in Sec. 3.9.

Generate potential solutions This step uses the data and various means of display to start to generate any solution that might provide an answer to the problem. The brainstorming technique described in Sec. 2.3 can be used for this, or any means of creating potential solutions. It is best to allow suggestions for solutions to flow freely without any criticism or evaluation. Grouping potential solutions is also useful as one solution may have attributes that could be applied to another, so making it stronger.

It is a good idea to generate as many solutions as possible. Sometimes it is useful to define the boundaries associated with the potential solutions to the problem. For instance, what things are you unable to do to solve the problem? Examining this boundary may yield another avenue to explore. What things do you not know that you would like to know? This may identify some areas for further data collection or analysis. What things are out of your control? Again this may direct the problem solver to a new train of thought which will help to solve the problem.

Another path to take is to look at previous examples of similar problems and see how they were solved. In their solutions, did they have access to the technologies and materials that you have? Consider the effective actions used in the past as potential solutions to your problem.

Take one word of warning. Sometimes a solution will jump out and the general feeling will be 'Why didn't we think of that before?' Such solutions may contain hidden pitfalls or traps and it is recommended that the team be very cautious about solutions that seem obvious.

Finally, it is often a good idea to involve people in the search for solutions who are not too close to the problem. This brings a fresh approach to the process and although these people may not come up with the ideal solution, they may act as a catalyst for the process and spark a new idea or new track to explore. The output from this step is essentially alternatives and one should seek as many of these as possible to ensure that all avenues have been explored. Some alternatives will be eliminated quickly, but this will serve to ensure that no stone has been left unturned and a complete and thorough investigation has been carried out.

Select and plan a solution Depending on the complexity of the problem and the solution, a selection method can range from a simple consensus to following a process of selection such as Pugh or Combinex (see Sec. 2.2). However, there are a number of areas that have to be considered before getting into the process of selection.

Firstly, what are the criteria for the selection of a solution? It is important to establish the extent to which the implementation of the solution is within the control of the team. Also important is to consider the degree to which the solution satisfies the requirements of solving the problem. Identify also how much resource (cost and manpower) can be expended in implementing any solution. Do a cost–benefit analysis to evaluate what is the expected payoff in implementing a particular solution. A judgement about the relative length of time that it will take to resolve the problem is another important criterion to consider. Once the solution to the problem has been implemented how well will it be accepted by the customer or the organization affected? All these questions should be addressed before the process of selecting a solution begins so that proper criteria for selection can be put in place.

Once a solution is proposed it may be worth considering breaking down the actions into smaller steps so that each step can be tested in a short trial. This would apply if the problem and/or the solution is particularly complicated, or if the outcome of the solution has a high degree of uncertainty associated with it. Sometimes the idea of breaking up the activity enables one to evaluate the impact of each stage of the solution and reduces both the risk and the cost of implementation. Also, any practical example that can be cited as a solution will enable the idea to be sold more easily to management. Once a course for improvement has been selected it should be turned into an action plan so that all the necessary resources can be brought into the action to effect the optimum sequence of events to implement the solution.

Implement the solution The implementation of a solution can only be described in very generic terms because the activities during this phase are very dependent on the specific problem and solution. There are, however, guidelines that should be adhered to during this phase of the process:

1. Do not take anything for granted. Question everything in the process. Sometimes there is a reluctance to ask questions because they seem trivial, obvious or stupid. There is a saying that the only stupid questions are those that are never asked! Encourage everyone in the team to ask whatever question comes to mind.
2. Ensure that everyone in the team knows what they have to do in the process of implementing a solution. If necessary assign roles and responsibilities and encourage the questioning of who has responsibility for the boundaries. We have all heard of things falling between the cracks. Make sure that there are no cracks. Make sure also that everyone affected by the solution is involved in implementing it. Nothing is worse than a situation where a solution has been fully implemented only to be scuttled by someone who was not involved and who comes along at the eleventh hour with an additional difficulty.
3. Ideally divide the solution into sequential and manageable steps that can be readily monitored.

Evaluate the solution One of the most neglected parts of any problem solving process is probably that of confirming that the solution has actually worked. This is important from two aspects. Firstly, implementation is not the conclusion to the process. It is important to have some feedback that the changes that have been made in an attempt to solve the problem have been successful. Secondly, it is quite unusual to implement a solution that solves the problem without creating some other (hopefully minor) problem in the process. Therefore any evaluation of the effectiveness of the solution is significant in case there are side effects that may need further attention.

Evaluation is done by the proper use of analysis tools and by good data collection. The team should not lose sight of the original problem and should continually ask, 'Did the solution effectively solve the problem?'

3.9 SIX TOOLS FOR THE DESIGNER

1. Brainstorming
2. Fishbone diagrams
3. Pareto diagrams
4. Histograms
5. Control charts
6. Force field diagrams

It has been said that a picture paints a thousand words, and this is no less true in the complex world of engineering where the designer and engineer are constantly wrestling with complex relationships and interactions. Most engineers and designers have some difficulty expressing themselves without being able to draw a diagram to assist them. This is possibly due to the fact that we spend a good deal of time trying to explain scientific and engineering principles and phenomena by drawing a picture to represent it.

The designer can use many tools of this sort to help a normal understanding of the design. Most of these tools are not difficult to understand beyond that of normal graphical representation, but the benefits that may be gained from employing these tools at the right time is enormous.

The six tools in particular that I have highlighted here are based on my experience. Timely use of them will certainly assist understanding. It may enable the breakthrough for which one searches.

Brainstorming

This technique has been dealt with extensively in Sec. 2.3. It is used during any period where innovation or invention is required, whether to find ideas for an original design or to find potential solutions to problems. What brainstorming can achieve is a large number of ideas in a very short time. As the ideas are generated the very process of generating them in a group atmosphere enables stimulation of ideas between contributors. Even the silliest idea is not wasted in this context as it can very well serve to trigger an idea from another member of the group which may turn out to be very valuable indeed.

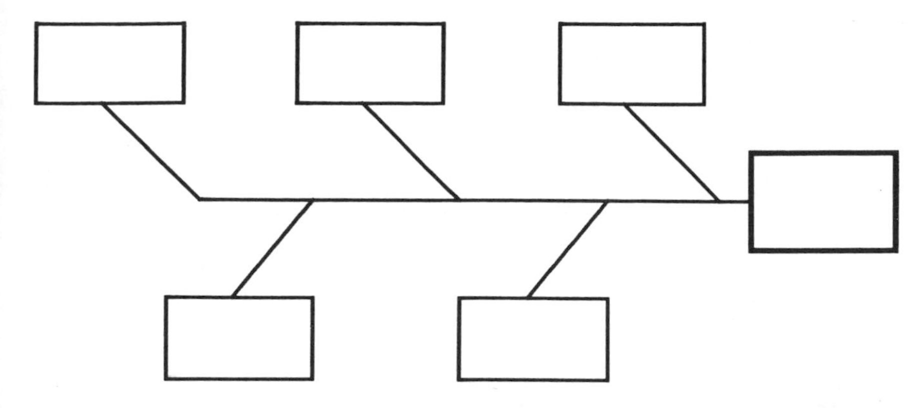

Figure 3.46 A fishbone diagram.

There are certain rules that should be obeyed and these should be imposed strictly. Firstly, there should be no criticism as the ideas are generated. There should also be no evaluation of the ideas. The group should aim to keep the flow of ideas going so that many avenues are opened up. They should aim to achieve quantity, not quality, and they should be encouraged to combine ideas or group them together part way through the process and certainly at the end.

Once a list has been generated this grouping will serve to focus the group's thinking on to a variety of categories for evaluation.

Fishbone Diagrams

The fishbone diagram is a graphical aid for grouping and organizing items that interrelate to deliver an output (Fig. 3.46). For instance, following a brainstorming session to generate possible causes of a problem, all those that are primary causes can be grouped on to one 'bone' of the fish and all those of another primary cause can be grouped on to another. Such primary causes might be people related, machine related, materials related or process related (Fig. 3.47).

The benefits of using such a diagram are firstly that it organizes thoughts into such a way that the designer can formulate a clear picture of things like relative importance of different parts of the fishbone. The designer can ascertain whether there are repetitions in different areas of the diagram and decide such things as where best to deploy resources to the situation.

Pareto Diagrams

A Pareto diagram (Fig. 3.48), named after Vilfredo Pareto, a nineteenth-century Italian economist who worked with unequal distributions, uses vertical bars to depict the value of things but arranges the bars in order such

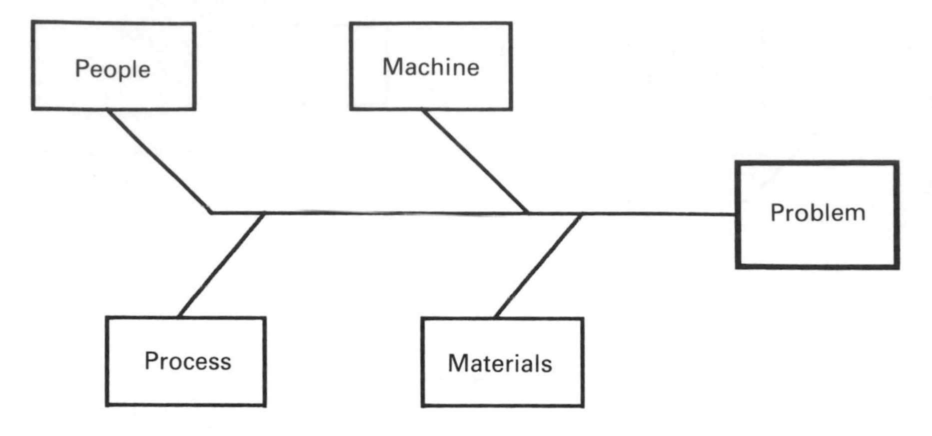

Figure 3.47 Primary causes on a fishbone.

that the bar with the greatest value is first, the one with the next largest value is second, etc. The largest value bars are placed on the left-hand side of the diagram. Essentially the diagram separates out the important few from the trivial many. This kind of diagram enables the designer to view the data and immediately get a graphical feeling about the relative levels of importance of

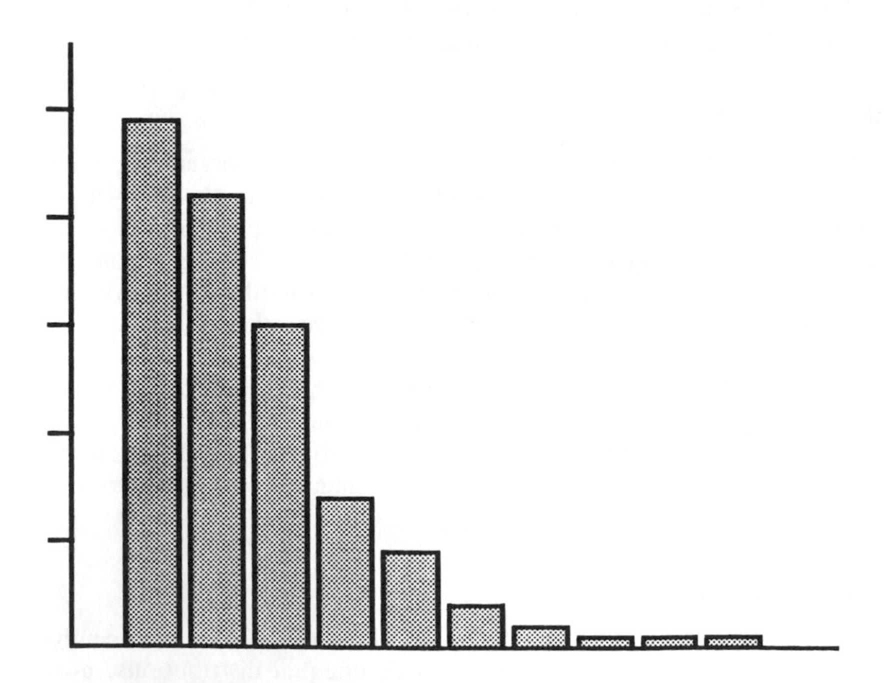

Figure 3.48 Pareto diagram.

each of the items. It is useful when deciding which element to work on first or which will provide the most benefit relative to the others.

It is often useful to take the one or two largest bars in the Pareto and break these down into further Paretos before deciding which elements to look at.

Histograms

A histogram (Fig. 3.49) enables us to graphically depict the amount of variability that lies within a process. It, like the Pareto diagram, uses bars to represent the value of things. These values are arranged on a horizontal axis which represents variability. For example, parts coming off a machine tool are measured for accuracy: 50 per cent are measured as 10 mm long, 25 are 9.8 mm long, 22 are 10.2 mm long, 2 are 9.6 mm long and 1 is 10.4 mm long. These are plotted as a histogram in Fig. 3.49 with the length on the horizontal axis.

A study of the spread and shape of the distribution tells us a lot about the condition of the process. Primarily, of course, one is looking for a spread that is within the upper and lower limits of the specification, and this can be

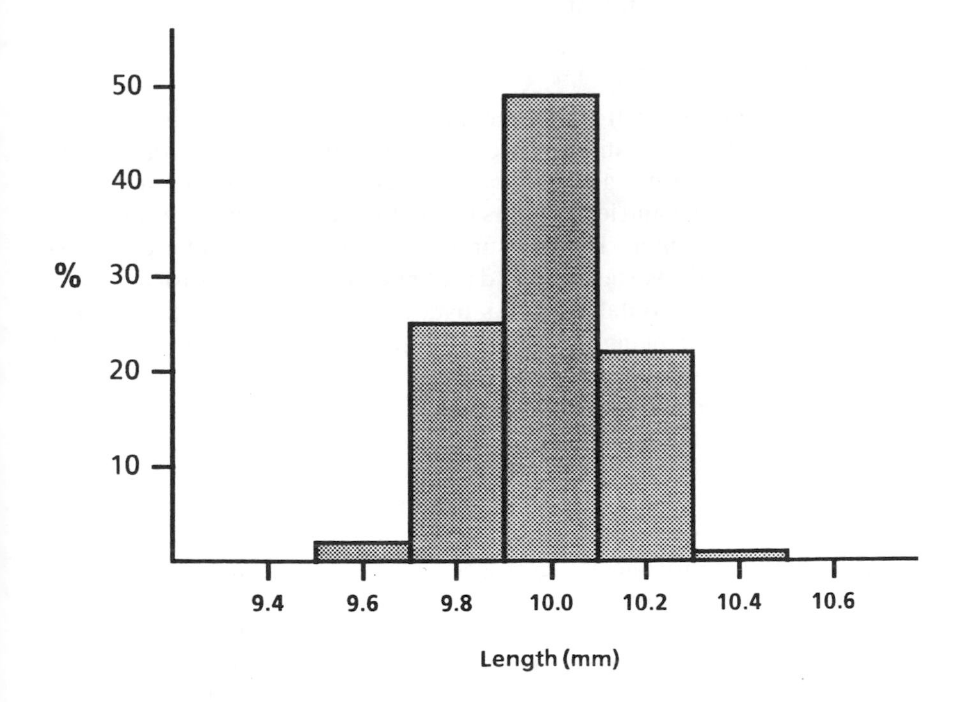

Figure 3.49 Histogram.

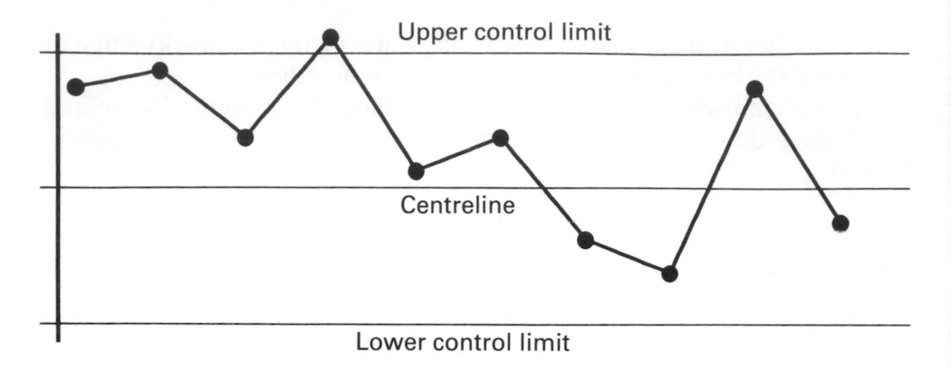

Figure 3.50 Control chart.

used as a process capability measurement (see Sec. 3.12). Obviously the narrower the distribution, the less the variability in the process.

Skewness of the distribution can also tell the designer something about what is happening within the process. Temperatures may be increasing as the process runs or wear may be taking place. Not all distributions are 'normal', that is bell shaped. Sometimes a distribution may have twin peaks (is bimodal). These may represent significant shifts in the process such as a different machine or different operator.

Control Charts

A control chart (Fig. 3.50) is another simple chart used to plot the progress of a particular factor within a process. The factor will be one of the key variables of the process and will be required to be kept within certain predetermined upper and lower limits in order to keep the process operating in a satisfactory manner. The measurement of the factor under consideration is plotted on the vertical axis and the upper and lower control limits are depicted as two horizontal lines. As data are collected from the process regarding the factor being measured it is entered on the chart using the horizontal axis as a time base.

The fluctuation of the points within the limits results from variation that is inherent within the process. This variation results from common causes built into the system and could only be affected by changing the system in some way. Points outside the limits come from changes that occur from special causes such as errors or freak conditions. In order to bring the process under control these occurrences must be eliminated, so enabling consistency within the process.

The control chart can be used to plot the distribution of the output from the process and assess its process capability (see Sec. 3.12). The charts are generally used to investigate the stability of a process, to bring it under control and to enable an assessment of the process capability index, C_{pk}.

Force Field Diagrams

A force field diagram (Fig. 3.51) is used whenever a change has occurred that creates a problem or whenever there is a desired change that needs to be characterized. The force field technique uses the fact that any change or potential change has two sets of factors influencing it: those that are driving it towards change and those that are restraining it from change. For example, take a desired change situation of someone who smokes but wants to give it up. Here there is a desired change of state from smoking to not smoking. What are the factors that drive this desire to change?

Reduced expenditure
Health threat

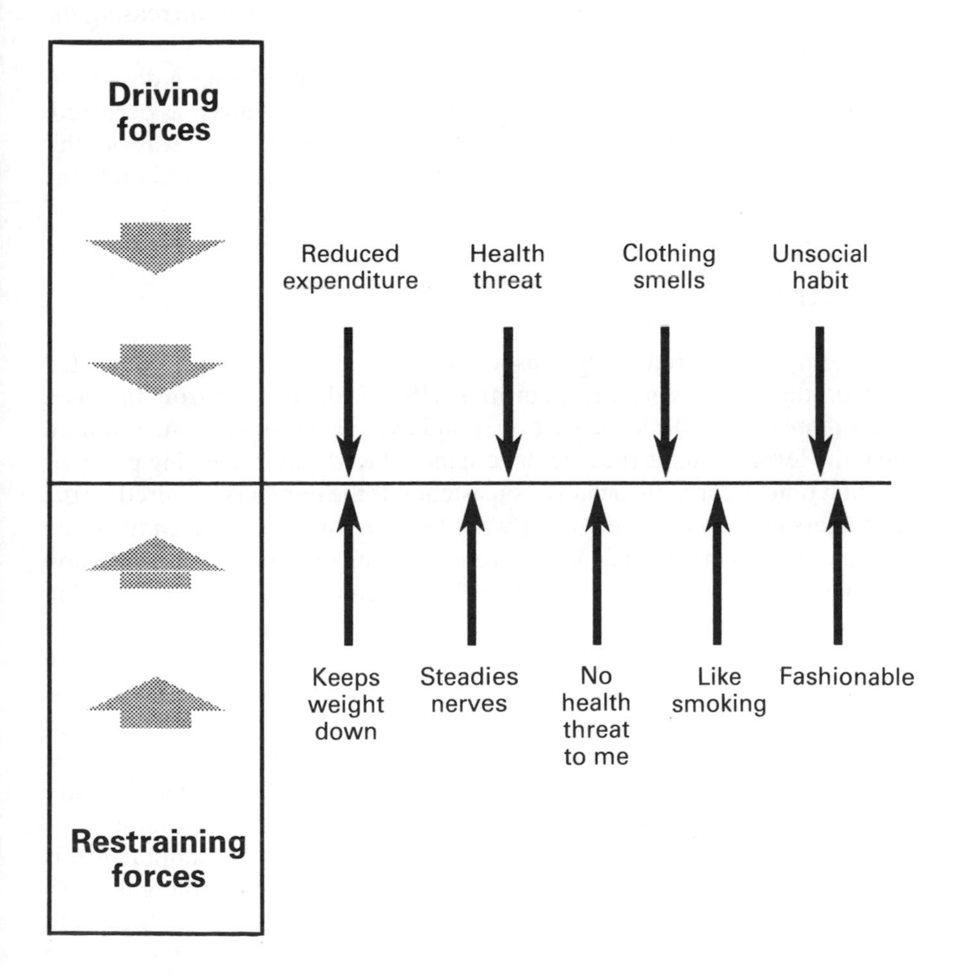

Figure 3.51 Force field diagram to stop smoking.

Unsocial habit
Clothing smells of smoke

What are the factors that resist the desire to change?

Like smoking
Fashionable
Don't believe there is a health threat to me
Steadies nerves
Keeps weight down

It will be noticed that if any of the factors that drive the change are increased, there will be a greater influence to change. If any of the factors that resist the change are reduced this also will generate a greater influence to change. Therefore if the change is desired we should concentrate on increasing the drivers and reducing the resistors.

A force field diagram can be used in this way when a change is desired. It can also be used in a similar way when an undesirable change has occurred. In this case the driving and resisting forces are listed in the same way but this time the user looks for factors that have increased or decreased and may have caused the change.

3.10 TOOLING

The term 'tooling' refers to an aspect of manufacturing that is connected with producing parts in large quantities. Historically it stems from the need to recreate exactly the same part over and over again to reach the required quantity level. Imagine the original engineer faced with recreating parts for the first time. In order to achieve consistency as the numbers required by the customers increased, the engineer would search for ways in which the same operations of manufacture could be repeated with only a minimum variation throughout the whole process. The first attempts would be to develop devices that would enable hand operations to become more consistent; we would call these jigs or fixtures. Later as processes became more and more varied and complex an equivalent complexity and variety of devices was required. These include devices to ensure accurate drilling of holes, both in terms of absolute position and relative positions; moulds to form the intricate shapes of plastic or cast metal and assembly aids that would ensure that parts which were to be joined together would also be consistent.

This whole array of devices to provide consistency in manufacture is broadly classified as tools or tooling. The consideration of tooling is a key factor in considering design. For instance, different tooling methods are necessary depending on the quantities of components to be manufactured. Any design intended to provide just 'one off' of a part or product can be manufactured by attending to each dimension individually. The mating parts can be 'fitted' so that they form a compatible pair, knowing that one half of the pair need not be interchanged with another part. The design therefore does not need to pay attention to absolute consistency of dimensions. As long as two mating dimensions are mutually compatible, there is no need for further control. Tooling for such a design is therefore unnecessary except as an aid to manufacture. Even here it may be redundant, since, if you can make the tool, you can make the part. Advantages of tooling are:

Consistent quality
Speed of manufacture
Better process control

Disadvantages are:

Lead time
Capital cost

Different tooling methods may cause the designer to change the overall design approach and therefore it follows that the quantities to be manufactured feature significantly in the design thinking. Of course tool design is a whole area of design expertise that demands knowledge of the manufacturing processes and that has its own limitations and constraints. The designer of a part must therefore be cognizant of the factors affecting tool design in order to make the part compatible with this process. In the product design process tooling must be a major consideration throughout all the design processes. I have said before that for a product the design must reflect the production intent design and once this is the case tooling becomes a significant factor. One of the problems in going through the various iterations of the design process is that the final design of tooling cannot be implemented until late in the process because, until the design has been completely tested and proven, the product team cannot commit itself to the high cost that the final tooling will incur. Additionally, there is the complication that this final tooling will not only be costly but will also need a great deal of time to prepare. So the dilemma is created that one needs to know early what the final design will be in order to start the preparation of the tooling, but one cannot start this until the design has been fully tested and confirmed as final. Now the question arises of how we can test a representative set of parts which, by definition, must be made from the final tooling when we cannot afford the cost of doing so until we have the answers from that test. The answer is that we cannot.

What we have to search for is a method by which we can make parts more cheaply using tooling that is a satisfactory representation of the final tooling. This is often called 'soft tooling', the final version being called 'hard tooling'. These two phrases are derived, I believe, from tooling used for casting or moulding. In this case the tooling made from steel was left soft until the first parts were run off and proved to be satisfactory. Once this had been completed the tool was hardened so that its wear rate and therefore its life could be extended enormously from the original 'soft' version. More generally, the two terms are used to describe the different tooling used for the early test models as opposed to that used for the production parts which will be made in large quantities.

If the design is intended to produce a product in batch production, that is a limited number of products made all together in a batch, the prospect is entirely different. Typically a batch may consist of perhaps only 100 sets of parts which will be assembled into the product at the same time. Now the manufacturer is faced with the requirement to have each mating part compatible with each of its 100 mates. This requires both the design and the manufacturing processes to be capable of producing parts that are accurate enough such that any set selected at random from the total will fit together. The manufacturing processes will in this case require tooling of a type that will last for the full 100 sets and the design will need to be done in a way that supports this type of tooling. A simple example of this is when one requires to make a number of lengths of metal or wood all the same length. If one uses the first to measure off the second and the second to measure off the third, etc., then the length dimension will tend to grow. If, however, one uses the first length to measure all the rest, then one is using this essentially as a 'tool' for the manufacture of all the subsequent pieces, and these will be almost exactly the same as each other.

In mass production, the process takes on more complexity. Mass production quantities range in the region of hundreds of thousands to millions, so here tooling is very important and has a number of other features. Tools will either need to last the full extent of the continuous production run, or be replaceable once they wear out. Wear-out is a problem since as tools begin to wear the dimensions that need to be controlled will gradually change. The production quantities may be so large that one tool may not be enough to sustain the production rate. Here duplicate tool or tools making multiple parts may be required. This is the responsibility of the production engineer to decide how many and in what way the tooling strategy will support the manufacturing strategy. Mass production tools will be very expensive. Many of them will be complex and consequently will have a very long lead time. Here it is important that the tools are able to be changed between the time when the work starts on making them until the design is frozen. Conversely, of course, it is a major objective of the designer to ensure that the design of parts for which tools are being prepared are subjected to minimum change. This in turn demands a full understanding of the design function and the development of a robust design.

Mass production has other differences with lower quantity manufacturing methods. Mass production requires the full coordination of parts, assembly processes, tooling, processes and people to be effective in order for the technique to work. In these specialist areas of manufacturing it is important to emphasize that the design process and the choice and implementation of the production process must be closely linked at the outset. Figure 3.52 shows a typical schedule for the development of tools and designs to provide parts for production quantities from hard tooling. It can be readily seen that the process has so many elements to it that the timing and planning of this activity must be closely controlled and monitored. Let us look at some of the elements of the process and see how they affect the design and the approach that the designer must take.

One of the major problems with designing for manufacturing tooling is that the time taken to produce tools is so long that the designer must be well ahead in the parts that need complex tooling. This is a problem, because the very parts that need complex tooling are the ones (almost by definition) that cannot be designed quickly. This is because they require enormous amounts of data about the other parts of the design to be available before the design can be finalized.

One aspect that would help this situation would be to reduce significantly the time needed to produce tools. Why do tools take so long to deliver? Here is an example of the various events that involve the delivery of a tool for high-volume production. Let us assume that the normal design, detail, check and sign-off of the drawing have all been completed. Following this point in the process the following steps are required:

Place order
Design tool
Manufacture tool
Introduce changes if required
Run sample parts
Review product of tool (parts)
Rework tool to correct errors
Review product of tool
Produce first production parts

This sequential process for the delivery of a complex tool has assumed only one rework to the tool itself. Quite often such a simple modification is not possible in practice and this therefore extends the lead time further.

3.11 TAGUCHI METHODOLOGY

Dr Genichi Taguchi has made a significant impact on the world of engineering by revolutionizing the way in which engineers think about their approach to design. Back in the fifties, when Dr Taguchi worked for the Electrical

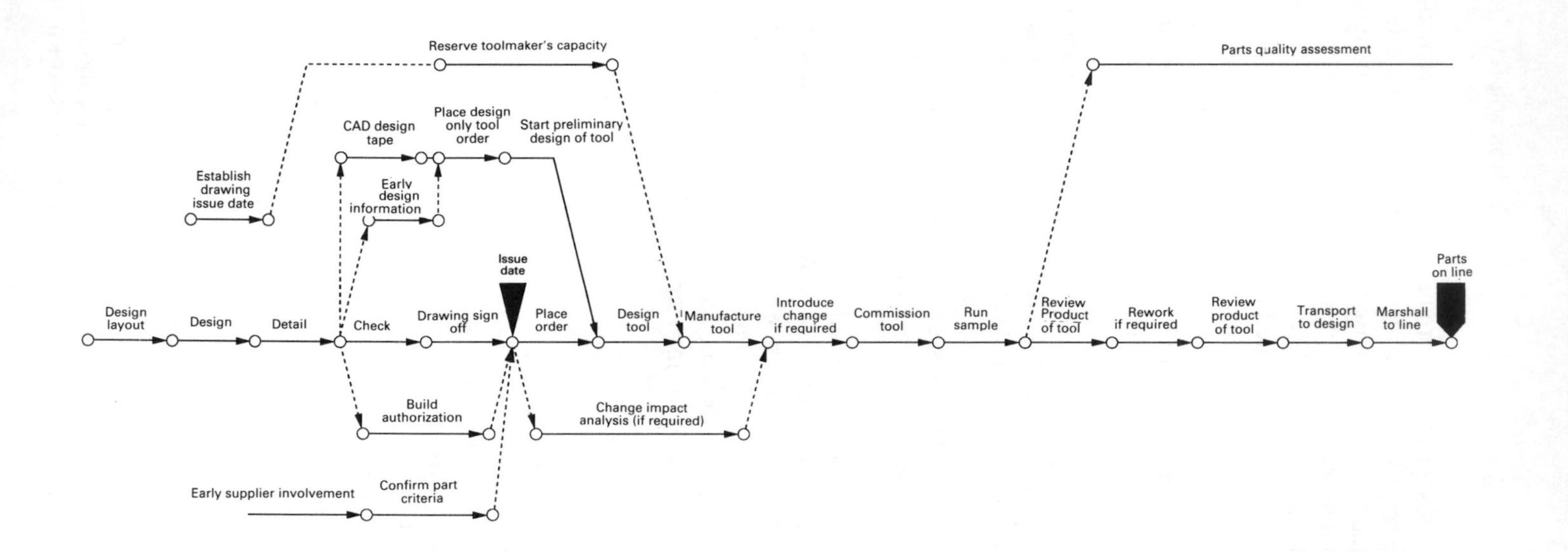

Figure 3.52 The plan for tooling.

Communication Laboratory, he recognized that in research and development most of the time and money was spent on experimentation and testing. His concept of an attack on each of the sources of quality deterioration to deliver quality engineering at low cost is called off-line quality control. It involves three major elements (Fig. 3.53) through the design process. These are system design, parameter design and tolerance design.

Systems design is aimed at finding the basic elements or methods to produce the required output. It involves innovation and requires knowledge of science and engineering in order to create choices of materials and processes that will achieve the best combination. The systems design also includes an early assessment of key inputs and outputs required to meet the product goals and, using these, an estimate of how well the chosen technologies will achieve the requirements.

Parameter design is the natural next step in the process which takes the systems parameters and determines the best values of these. It is the process of engineering that develops the best settings of each of the parameters to enable the required output to be on target and have low variation. The changes in all environmental conditions, or noise factors, must be taken into account at this stage and it is this process that enables the design to be robust.

Tolerance design deals with the process of identifying the quality sensitive components and putting limits on these values to meet the required level of variation of the output.

The following table shows the estimated percentage of time spent on each of these parts of the design in Japan as compared with the Western world:

	US	Japan
System design	70%	40%
Parameter design	2%	40%
Tolerance design	28%	20%

It seems that most engineers in the United States step immediately from systems design to tolerance design. This of course costs money in terms of better materials, components or the processes to provide them.

Taguchi methodology is centred around the quality loss function (Fig. 3.54). Every parameter in a design will have a target value and upper and lower limits around that target. Assuming the target value to be based upon

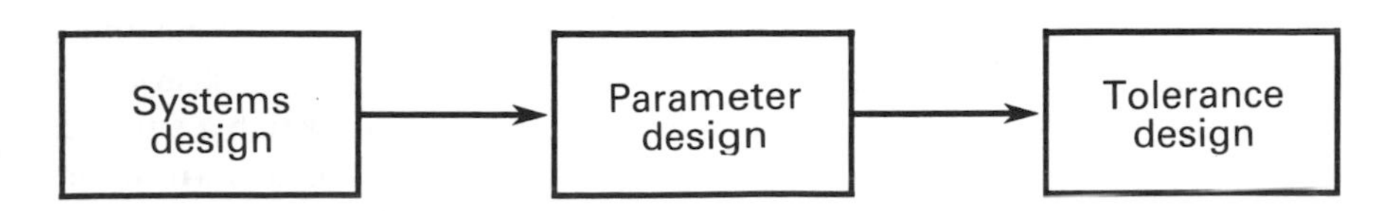

Figure 3.53 The Taguchi process.

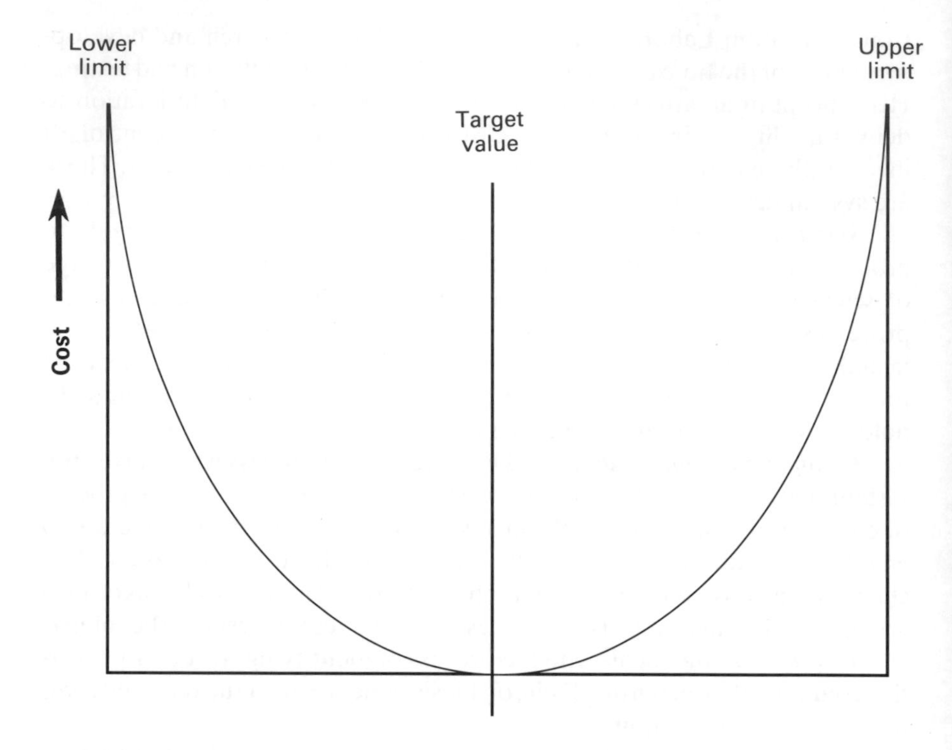

Figure 3.54 The quality loss function.

the desire to meet customer satisfaction, any deviation from that value will mean a level of reduced satisfaction for the customer. Furthermore, the further away from the target value the parameter is, the greater will be the dissatisfaction to the customer. If one plots this on a scale of cost of exceeding the specification, the curve looks like that shown in Fig. 3.54. The concept shows that it is not acceptable just to keep the parameter within the set limits, but that it is necessary to keep as close as possible to the nominal or target value.

Variability and the Quality Loss Function

When a product is delivered to the customer the extent to which it meets the specification defined by the supplier will determine the level of customer dissatisfaction with that product, that is assuming that the customer's requirements have been correctly translated. One way in which a supplier might make all the customers happy is to supply all the products within the boundaries of the lower and upper specification limits (Fig. 3.55).

This is a very coarse picture of the true situation, because there may be

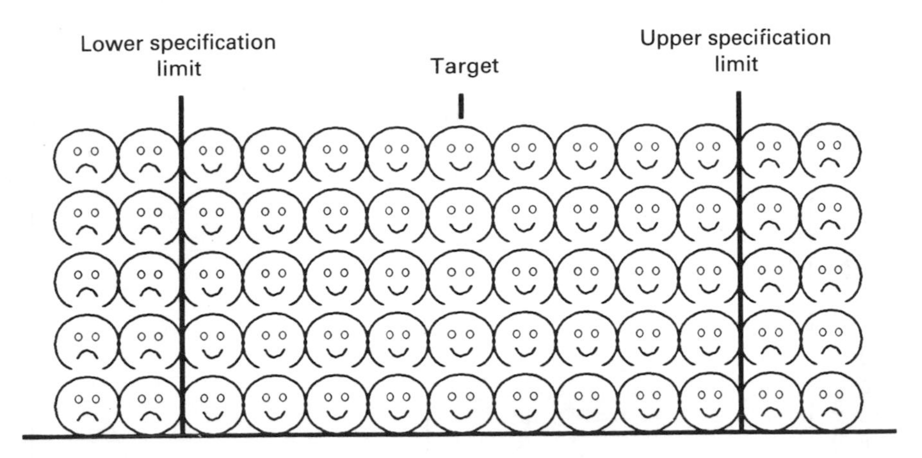

Figure 3.55 Happy customers with products within the specification boundaries?

some customers who are actually unhappy if they get a product close to the specification limits. Likewise, some customers may be satisfied when a product is just outside the specification limits. The only true reflection of customer satisfaction is to say that the proportion of dissatisfied customers increases as the product performance moves further from the target operating point towards either the upper or the lower specification limit (Fig. 3.56).

The value of this hypothesis lies in the fact that the vertical scale of the diagram can be converted to cost. Thus one can say that the loss of quality can be depicted by the curve $L = ky^2$, where y is the variable that the customer assesses (or response variable), and L is the cost incurred in

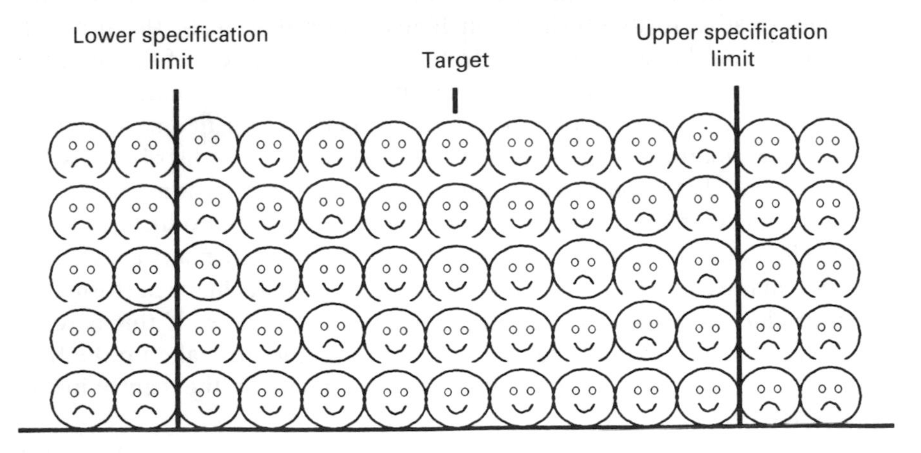

Figure 3.56 The true picture of happy customers.

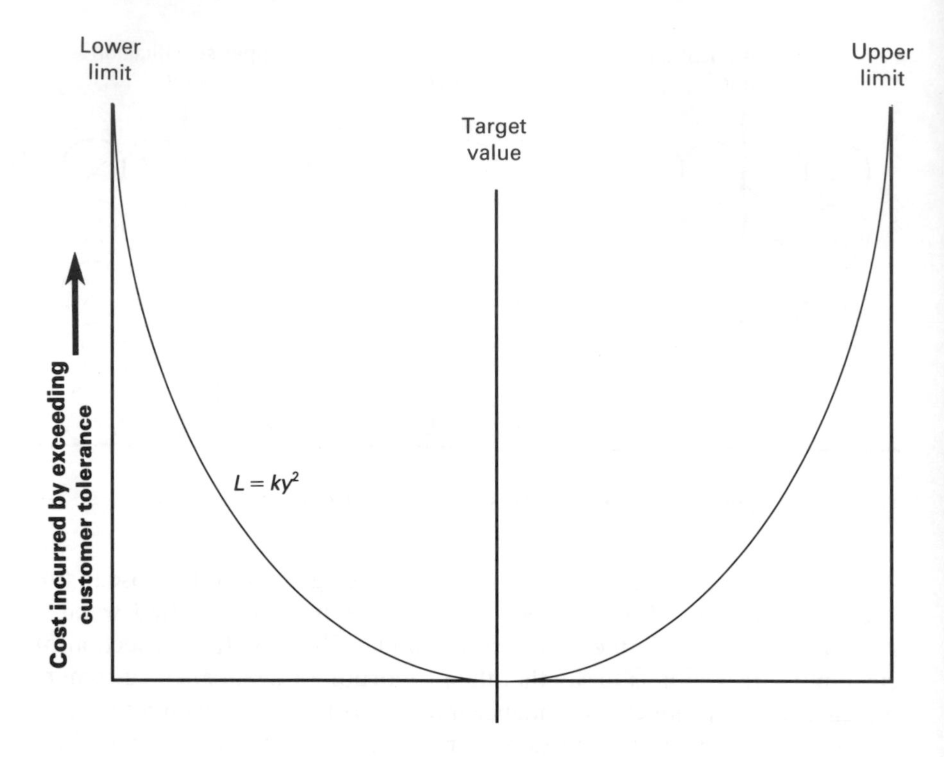

Figure 3.57 Cost incurred by exceeding customer tolerance.

exceeding customer tolerance and k is a proportionality constant. The curve $L = ky^2$ is called the quality loss function (Fig. 3.57).

Consider the case of a single product built at two different manufacturing sites. In the first case the distribution of the performance level of the product relative to its specification limits shows that all of the products operate within the defined limits. In the second case a few of the products are outside the specification limits but more are grouped around the target value (Fig. 3.58). If we look at the proportion of dissatisfied customers as defined by superimposing the quality loss function on the distributions, we will see that the second case creates less dissatisfied customers. This is because the vast proportion of these customers have products that are centred around the target value, whereas in the other case many are close to the boundaries of performance. It is very important to achieve performance of a product as close to the centreline target as possible. If this is to be achieved it means that all the other variables that go to make up the final performance of the product must also be aimed at the centreline of performance.

Taguchi has also introduced the concept of studying the signal-to-noise ratio when analysing data from a designed experiment. The signal-to-noise

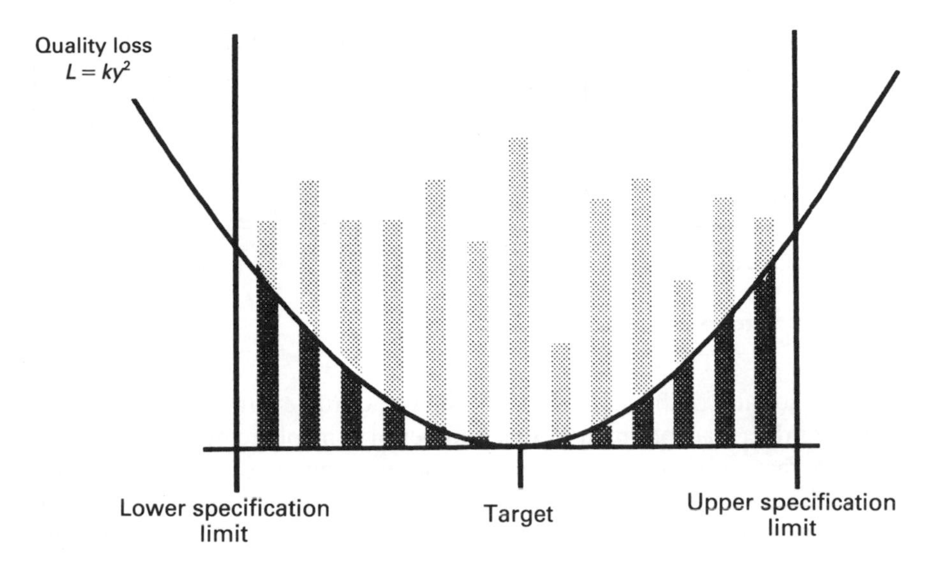

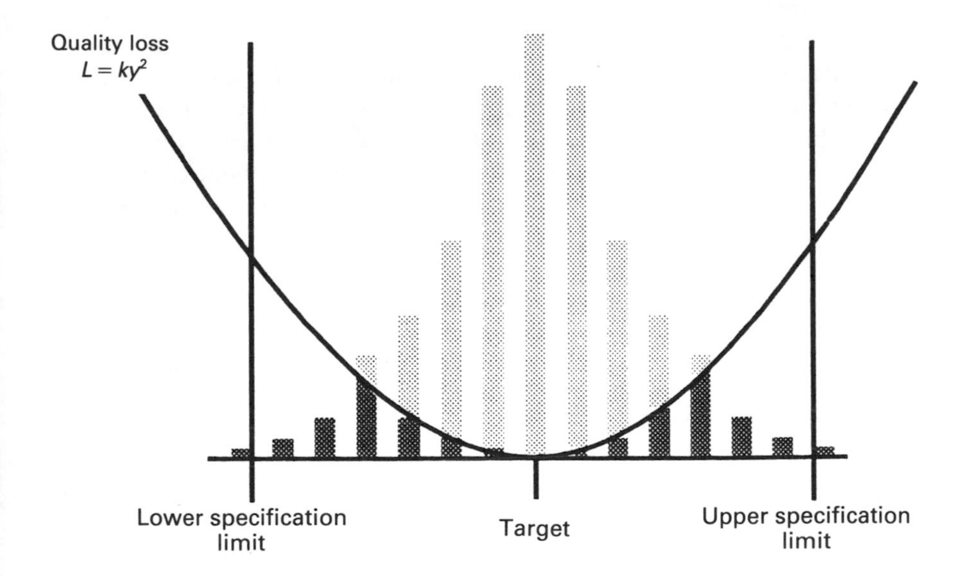

Figure 3.58 Performance level versus quality loss function.

Figure 3.59 Three cases of signal-to-noise ratio.

ratio is the ratio of the mean signal to the standard deviation (that is noise). If we conduct an experiment with a number of control factors and a number of noise factors, we can select the best combination of control factor values by analysing the signal-to-noise ratios. Three standard cases of signal-to-noise ratio are shown (Fig. 3.59): 'nominal is best', 'higher is better' and 'lower is better'.

Another important aspect of the Taguchi methodology is his approach to experimental design. Experimentation is not a new pursuit. Scientists and engineers have been experimenting for centuries. Often, however, the experimenter will be concerned with changing only one factor at a time. This is because of the apparent complexity of doing more than that and due to the danger of becoming 'out of control'. It is, however, possible to change more than one factor at a time and still maintain control. The methodology is called statistical experimental design. Basically, if we were to conduct an experiment using all factors, each having a high and low level, the number of runs increases enormously as the number of factors increases. Alternatively we can use a fractional factorial experiment which cuts down the number of runs significantly. Taguchi propounds the theory of using orthogonality, that is the ability to study each factor independently without confounding relationships. In other words, the variability of the factors are considered to be orthogonal to one another so that changes in one factor have no effect on another. In using an orthogonal array the experimenter can reduce the time taken to run the total experiment without sacrificing the resulting quality.

3.12 PROCESS CAPABILITY

When manufacturing engages in the process of making the part for the final product it is essential to be sure that the parts being produced by the manufacturing process are meeting the requirements set out by design. The idea of the manufacturing of parts being an activity remote from design has been fairly well discounted both in this book and by industry in general. No large scale manufacturing operation can function successfully to produce quality parts without a strong linkage with the design operation throughout the design process. The two processes must be complementary to each other, and here the variance of the manufacturing process must be within the tolerance of the design. Refer back to Fig. 2.14.

How capable of matching the design tolerance is the manufacturing process that has been chosen? This question must be answered during design, so that manufacturability is assured, and also throughout manufacturing, so that process capability is maintained.

Let us look at how we quantify and measure each of the two constraints. Design tolerance is defined by the drawing according to a critical specification that is generated by a critical parameter, that is a parameter that is a key part of the functional operation of the part or system. The tolerance will be defined ideally by a nominal value and a plus and minus tolerance. The tolerance will literally reflect the deviation from the mean which can be tolerated without a failure being induced. For a dimension that has a low sensitivity to the response to which it relates the tolerance may be wide. Conversely, a dimension that is highly sensitive to the relevant response is likely to have a close tolerance. Whatever the tolerance, it represents a boundary that the designer has mandated shall not be exceeded if the part is to function in an acceptable way.

The job of the manufacturing part of the team is to use a process to manufacture the part that will achieve the designer's requirements throughout the production process and when subjected to all the influences on the production process that cause variation in the final product. Most production processes when in control deliver an output whose distribution is usually normal about the mean (Fig. 3.60). It is therefore desirable to have a design tolerance defined in this way so that the process capability can be readily measured. If one draws the normal distribution which represents the out-

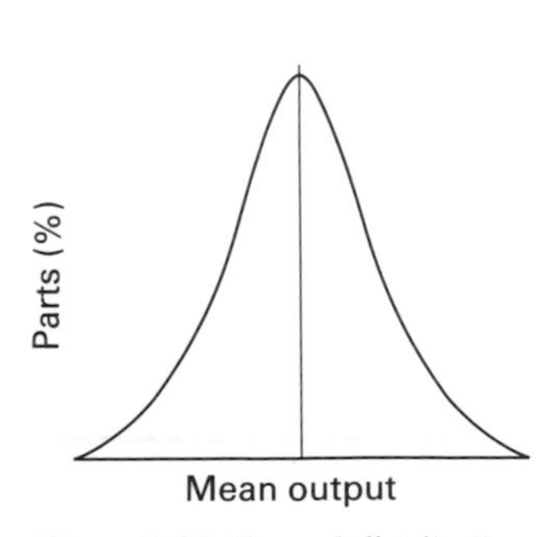

Figure 3.60 Normal distribution.

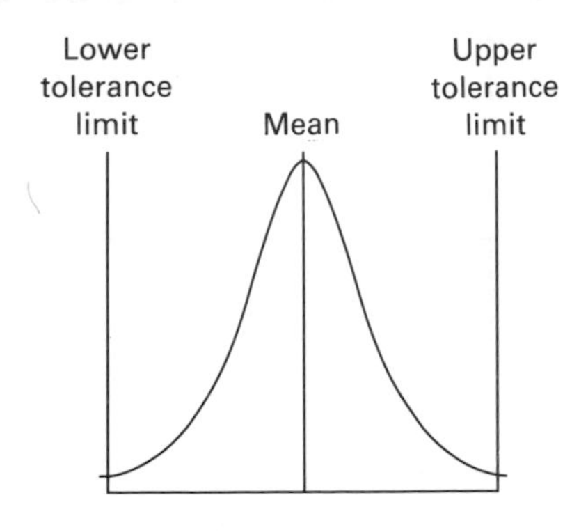

Figure 3.61 Distribution of design tolerance with upper and lower specification limits.

come of the manufacturing process, then one can also include on this diagram the design tolerance boundaries (Fig. 3.61).

Extremes of distributions are normally measured in terms of standard deviations (ς), and a 3-ς spread about each side of the mean of the distribution will show that 96.97 per cent of the parts are within the design tolerance boundaries. Similarly, a 6-ς variation about the mean would signify that 99.999 997 8 per cent of the parts would be inside the design tolerance.

In manufacturing it is usual to compare the distribution of the output to a 3-ς situation; this measurement is defined as the process capability index C_{pk}. Thus the C_{pk} value of a distribution that meets 3-ς tolerances is equal to 1, and one that meets a 4-ς variation has $C_{pk} = 1.33$. Thus a manufacturing process can be quantified in terms of its process capability by evaluating its C_{pk} number.

How do we calculate the C_{pk} number? There are two essential steps to evaluating a C_{pk} number. Firstly, the manufacturing process must be monitored and a control chart produced showing the variation of the process between the specified upper control level and the lower control level. This gives the data over a period of time needed to establish the stability of the process (Fig. 3.62). Secondly, the variation of this process must be verified as satisfying a normal distribution (Fig. 3.63).

Once these two steps have been completed, the data can be plotted as a distribution and compared with the design tolerances. Using standard formulae the C_{pk} number can be calculated (See page 176).

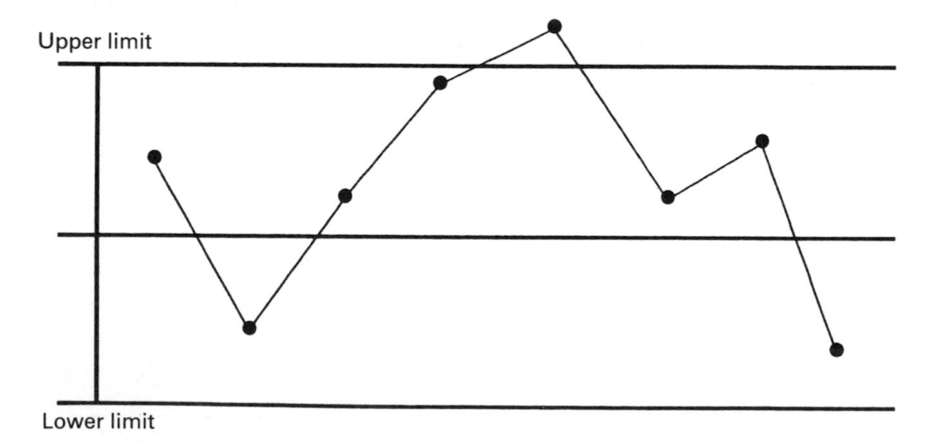

Figure 3.62 Control chart data.

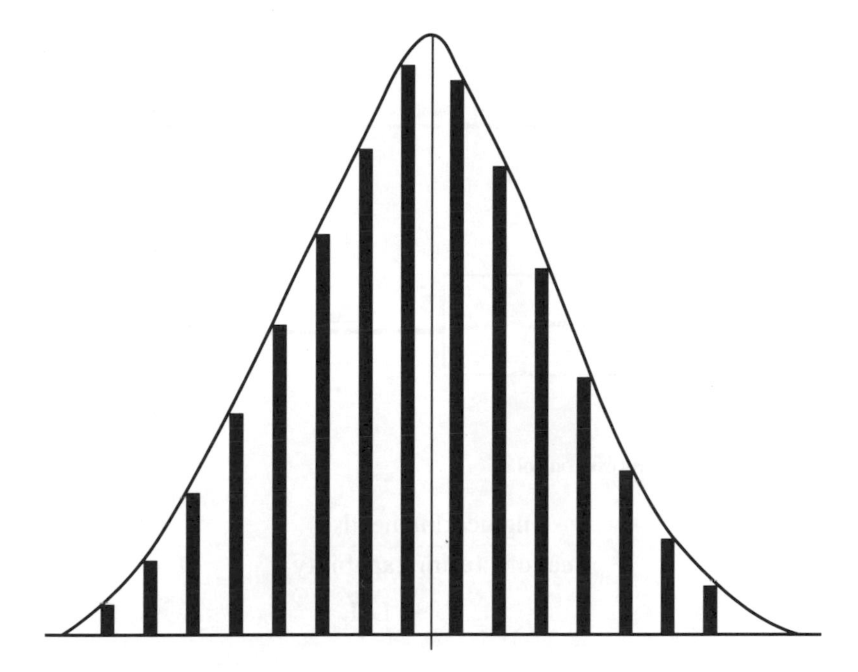

Figure 3.63 Fit to normal distribution.

3.13 STATISTICAL PROCESS CONTROL

When considering the manufacturing process and its measurement it is important to understand that:

No two things are produced exactly alike.
Things usually vary according to a pattern.
The shape of the distribution of parts produced by any process can be determined.
The shape of the distribution is distorted by variations due to assignable causes.

Statistical process control (SPC) is an activity that uses *statistical* methods to evaluate the *process* of making parts and enables the application of *control* of that process. It applies to the improvement of future parts as opposed to past or present parts, which can either be used as they are, scrapped or reworked to make them acceptable.

SPC links the output of the manufacturing process to the desires of the customer and feeds back information about the output to enable the process to be changed so that future output can meet the customer's needs (Fig. 3.64). The measurement tool for this is the process capability index, C_{pk}. This index compares the requirements of engineering with the capability of manufacturing and is a ratio,

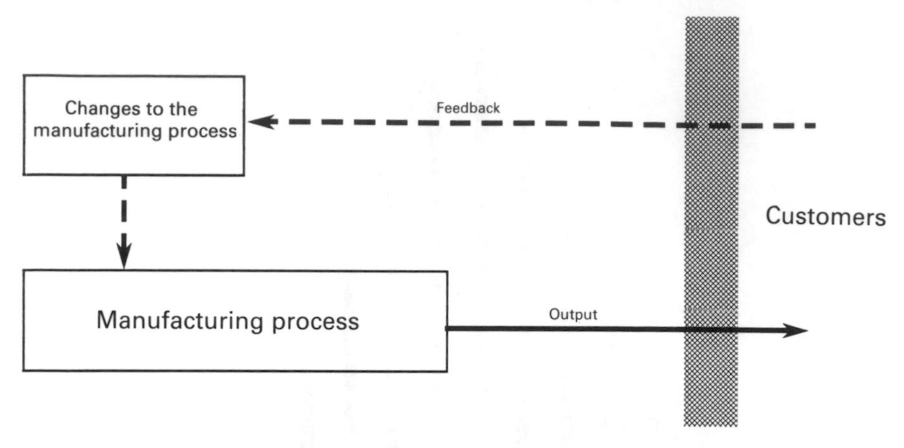

Figure 3.64 Statistical process control.

$$\frac{\text{engineering needs}}{\text{manufacturing capability}}$$

$$C_{pk} = \frac{x - \text{LSL}}{3\varsigma}$$

$$\text{or} \quad \frac{\text{USL} - x}{3\varsigma}$$

Take, for example, the size of a hole in a part whose design nominal is 20 mm and whose design tolerance is ± 2 mm. A distribution of these holes taken from the manufacturing process has a mean of 19.5 mm and a sigma of 0.47. For the process,

$$C_{pk} = \frac{19.5 - 18}{3 \times 0.47} = \frac{19.5 - 18}{1.41} = 1.06$$

$$\text{or} = \frac{22 - 19.5}{3 \times 0.47} = \frac{22 - 19.5}{1.41} = 1.77$$

Obviously the higher the value of the ratio C_{pk}, the better is the capability of the process. In this case the worst process capability index is 1.06. Most manufacturing plant will have targets to be met to establish a satisfactory process capability.

The steps in the SPC process are:

1. Make a choice of the parameters that are to be measured. This choice should reflect the critical parameters that have been defined during the design of the product. It is these parameters that will have a major impact on the performance of the product and therefore on customer satisfaction. The chosen parameters should also be measurable.
2. From a large number of parts taken from the current manufacturing process that is to be assessed, a histogram is generated and information

about the distribution of the chosen parameters of the parts can be recorded. The mean value of the data and the value for sigma (ς) are the two values required to calculate C_{pk} for the process. Note that this step is the only step where the engineering specification is compared with the capability of the process. Only if this step is fully satisfied in terms of meeting the desired tolerances does SPC move on to the next step. If there are unacceptable results from step 2, an analysis must be initiated to understand why this is so.

3. Samples of five parts are taken at a variety of times during the production run. Charts are prepared to identify the ranges of the process variation and the averages of the process. Using this information the quality engineer can tell whether the process is in control or not.
4. The data from step 3 is used to correct the process where appropriate or to confirm that the process is under control.

Processes can sometimes be under control technically but out of control statistically, and vice versa. They can of course be out of control or under control both technically and statistically. To illustrate this Fig. 3.65 shows the kinds of distributions that might be observed for various conditions of control.

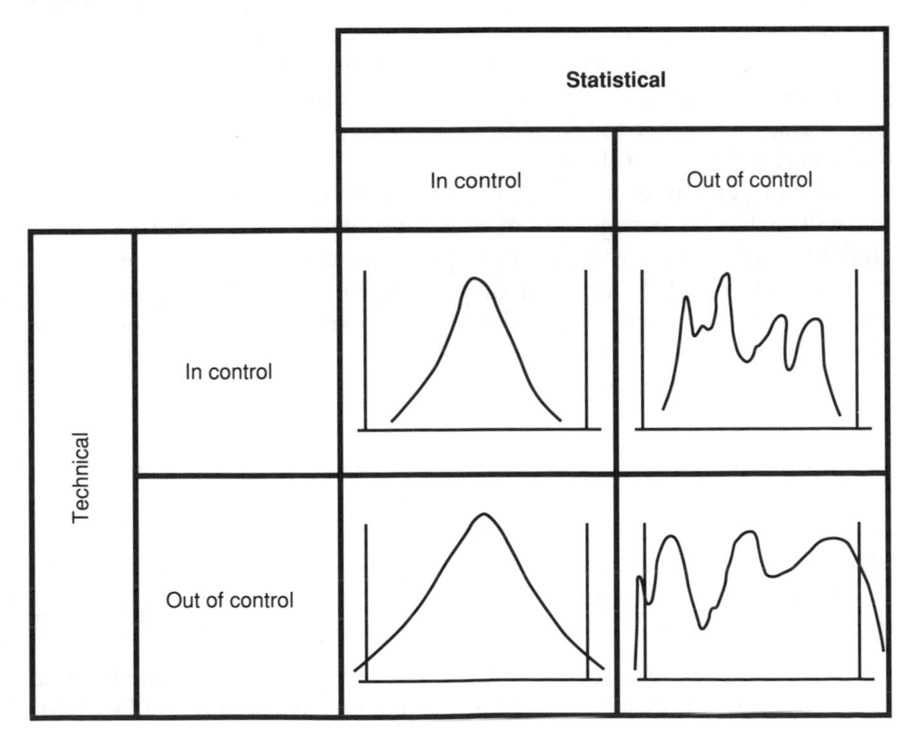

Figure 3.65 Technical and statistical process control.

In the lower right quadrant of the diagram the data shows a random form and is exceeding the technical specification limits. The process is therefore out of control both technically and statistically. In the upper right quadrant, although there is a similar distribution of data, showing statistical out-of-control, the data lies within the technical specification limits. In the lower left quadrant, a normal distribution shows good control of the process, but it exceeds the technical limits prescribed. Finally, in the upper left quadrant a normal distribution within the technical limits shows a process under control in both aspects.

3.14 PROBLEM MANAGEMENT

Much of the engineering associated with product development relates to the solving of problems. Engineers are often accused of being self-effacing or pessimistic because they tend to dwell on difficulties and problems rather than the successes of the job. This, I am sure, is true and may be one reason why the general status of the engineer is lowly compared with that of a doctor or a lawyer. Whichever way one looks at it, designing products has a large proportion of its emphasis on resolving the problems that arise during the process, and engineers and designers are trained to think and act in a way to solve these problems, and seem to enjoy doing just that. This is why it is important for designers and engineers to have skills in the process of problem solving and be able to take a direct route to potential solutions and take decisions on the best of these.

In management terms it is just as important to have a process in place that will enable the management of a set of problems within the programme. Only by efficient management of the problem set will the team be able to deal with the problems in order of priority, be able to identify the most important problems and apply the right proportion of resources for an optimum approach.

A good problem management system is one that

(a) Enables a problem set to be developed
(b) Provides a method by which the problems can be categorized and prioritized
(c) Provides a process for monitoring the progress of problem solving
(d) Enables the solution to be documented in a way that will inhibit the recurrence of the problem
(e) Assigns ownership to each of the problems encountered.

No doubt there are many problem management systems that achieve these objectives. The one described here is one that I have been involved with for many years and one that I have seen evolve to meet needs as they arose. I believe it will serve the average product development team admirably.

A Problem Management Process

Any problem has two important dimensions to it which need to be understood and defined by the problem solver. Firstly, how threatening is the problem to the success of the project? Could it cause the programme to be stopped if it is not resolved? Would the programme 'live with it' if the problem became impossible to solve? Secondly, how difficult is the problem to fix? Will it take more resources to get to grips with the problem or is it expected that the normal development process will yield a solution? The two questions are inextricably intertwined, as we shall see in a moment.

The threat that a problem poses to a project may be defined against one of three criteria:

1. The problem may be *critical*, that is it threatens the life of the programme if it is not solved. This may be a purely technical problem, such as one that threatens the safety of the user of the product, or it may be one of a business nature, such as a cost so high that the product becomes unviable. A 'critical' problem therefore is a 'showstopper' to the programme.
2. The problem may be *major*, that is it will have a significant effect on either the cost, quality or schedule of the product. Such a problem will not yield unless extraordinary measures are taken to resolve it. These may include extra resources or specialist input.
3. The problem may be *ordinary*, or its nature is such that it would be expected to be resolved by the normal process of events.

Each of these three categories defines the threat that the problem poses to the project in its fundamental form. How long will it take to resolve? A critical problem may be quite easy to resolve once the true definition of the problem is known. By comparison an ordinary problem may take a long time.

Categorization of problems in this way is related to the problem solving cycle (see Sec. 3.8). The process of solving problems is divided into six steps:

1. Definition of the problem
2. Analysis of the data
3. Search for potential solutions
4. Planning a solution
5. Implementing a solution
6. Evaluating the solution

This method of categorizing the progress of the problem through the problem solving cycle can be used to define the category of the problem in terms of its difficulty to reach a solution. For example, a problem that has just been discovered and considered to be a life threatening problem could be categorized as Critical/1, or C1. As the problem is processed and reaches the stage where a solution is being implemented the problem would be recategorized as Critical/5, or C5. Similarly, an ordinary problem fully analysed would be defined as Ordinary/2 or O2.

Using this method any set of problems can be categorized into the two important dimensions, that of criticality and that of progress in the problem solving cycle. This array of problem categories can then be used to both assess the health of the programme at any point in time and also to prioritize the problem solving activities.

The chart in Fig. 3.66 is a useful way of arranging the problem set and depicting those problems that should receive the most urgent attention. Thc problem management system must also enable the problem set to be developed, and must monitor and document the solutions. Firstly, it is important that problems are sought out by the programme team and not hidden. In fact it may be necessary to change the thinking of the team from one where success in avoiding problems is rewarded to one where success in finding problems is rewarded. There is no doubt that if problems are not found by the team they will be found by the customer in the field and will cost a tremendous amount more to resolve. There should therefore be an environment where the finding of problems is recognized as progress. At the same time it should also be emphasized that the rate of solving problems should at some stage be greater than the rate of finding problems.

A programme team can generate their problem list in a number of ways:

1. Problems identified from previous programmes
2. Problems identified from knowledge of the technology
3. Problems identified through the FMEA process

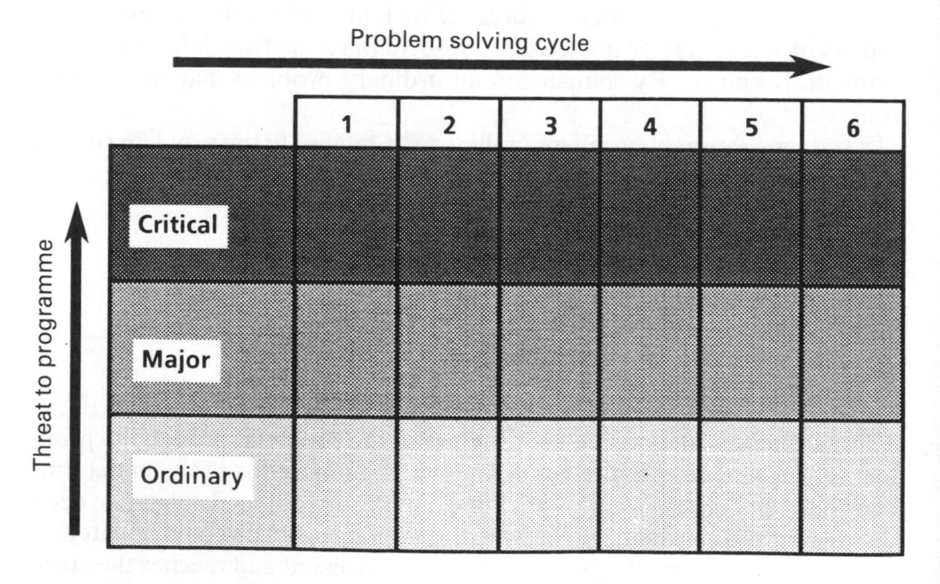

Figure 3.66 Problem management.

4. Problems identified from testing
5. Problems identified from hardware build

Generally the bulk of the problems are generated from the building or the testing of the hardware. It can be useful both to apply 'lessons learned' from previous programmes and also to study the history of the technology development as methods of predicting likely problem areas which should be kept in the mind of the designer.

Using these methods the problem list will contain a wide variety of problems, some major, some ordinary, some well defined and others only vaguely identified. The first job therefore is to categorize the problem set in the form described above, so that the criticality of the problem and the status of the process for solving it are both known.

Each problem on the problem list can then be identified as either critical, major or ordinary. Obviously any critical problems, which are defined as life threatening to the programme, must be dealt with urgently, probably under the direct management of the chief engineer or programme manager. The management of these problems may require the redeployment of some of the resources in order to bring sufficient attention and activity to their resolution.

Another important factor in the problem solving process is that each problem must be assigned 'an owner' who accepts responsibility for solving the problem. It is almost inevitable that any problem that does not have someone assigned specifically to it will either be neglected or overlooked. Ownership of problems is an essential aspect of the problem management process.

All problems on the list should be managed on a priority basis commensurate with the importance of the problem. For instance, all major problems would be reviewed by the chief engineer at a weekly problem management meeting where the problems listed would be addressed on a rota basis. Ordinary problems could be managed by the relevant subsystem manager and, as stated earlier, critical problems should be addressed immediately and continuously until resolved.

The process of problem solving has been described earlier, but it is useful to document the progress of this through a computer database.

CHAPTER

FOUR

MANAGEMENT PROCESSES

4.1 THE PRODUCT DEVELOPMENT CYCLE

The aim of this book is to provide the designer and engineer with a structured process that will enable them to deliver high quality designs. It deliberately avoids too much emphasis on management techniques as a means of enhancing design quality, but because everyone in the design team is a manager to some extent and because everyone is subjected to management as they proceed with their job, it is necessary to include a significant part covering the management process.

In order to put the management process in perspective, I refer back to Sec. 1.3 where the various elements of the design process are distinguished from each other: the management process, the design process, and tools and methods. Each of these should be regarded as distinctly separate from one another, and although each plays an important and individual role in producing a good design, they should not be confused.

The management process does not tell the designer how to do things differently. It does not provide any new vehicle to make designing more reliable. What it does do is provide a means for the programme to be monitored and controlled and for decisions regarding the business viability of the programme to be made. When carried out properly and with the right structure, it enables the programme to progress at the right times into the different phases needed for healthy development, while at the same time having the power to stop the programme should the business environment change or progress be damagingly inadequate.

The management process must not be served by the programme. The process itself must serve the programme and as such must contain the capability to be sufficiently flexible to accommodate the subtleties and variations of each particular programme.

The management process described here is comprehensive in that it covers the management of a programme from inception to end of life. Of course it may not be necessary to apply this level of detail and complexity to every programme, and for small products no doubt some of the detail can either be eliminated or tailored severely to suit.

The process structures the life cycle of the product into seven distinct phases. This process phasing establishes the decision points at which a review will take place by management and where approval control will be exercised. The life cycle begins with the inception of an idea or a statement of a marketing requirement. Technical goals are set for the product and against these an architectural framework and the technologies that will operate within this framework to deliver the requirements are selected. The cycle continues with the definition of a design to meet the customer's requirements and the building of production hardware. Launching of the product, asset management and servicing in the field are followed by the wind-down of the product and eventual withdrawal from the market-place (Fig. 4.1).

The seven phases of the management process are as follows:

1. *Preconcept phase.* This phase is initiated by the identification of a product need by the marketing group which is consistent with the business strategy of the company. It exists to validate the business opportunity by evaluating the product options against the business case itself, taking into account the competitive situation and the technological capabilities of the company at the time. Specifically it includes the selection of the architecture of the product, the technology set to meet customers' requirements and the definition of the agreed product goals between the marketing and the engineering groups. It culminates in the management decision to go ahead (or not), based on the ability of the technical community to meet the marketing needs.
2. *Concept phase.* This phase seeks to demonstrate that the technology set selected, working within the architecture selected, can meet the customers' requirements and the business case objectives. It includes the development of technologies, a production intent design and the establishment of the programme metrics which define quality, cost and delivery. Using these metrics the business case is assessed once again.
3. *Design phase.* This phase is used to complete the production intent design and confirm by demonstration of the hardware that the necessary quality, cost and delivery requirements can be met. The phase also includes the development of the plan to launch the product with the associated involvement of the sales, service and other support areas.
4. *Demonstration phase.* The intention during this phase is to confirm that the product design has the stability required to enable it to enter

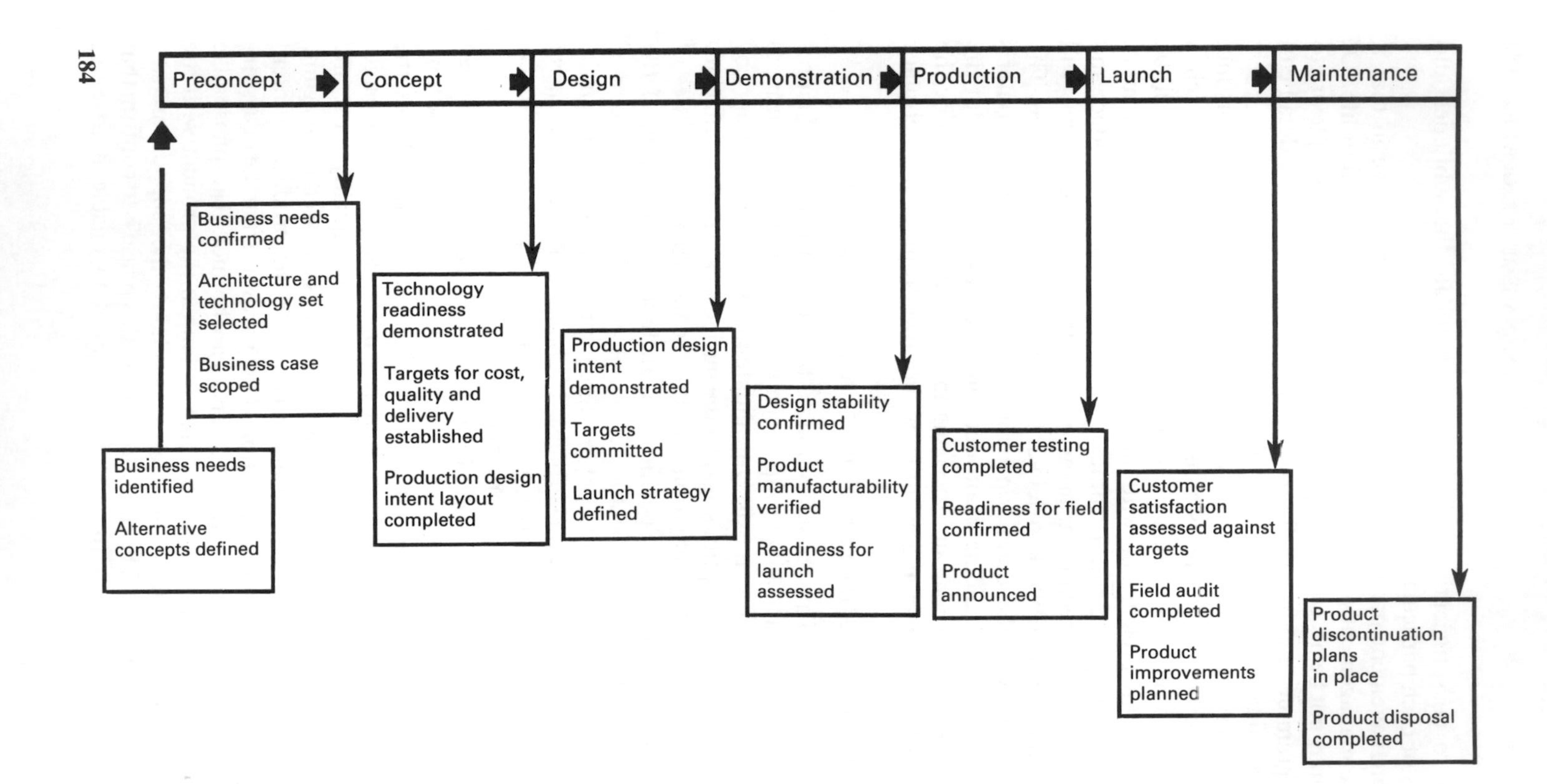

Figure 4.1 The product development cycle.

whatever production facility is planned for the product. This requires that the design and the hardware are compatible with the production and assembly processes that will be necessary to maintain the production levels planned. The phase authorizes the go-ahead of production materials procurement and initiates the necessary expenditure for production facilities to prepare for production scale-up.
5. *Production phase.* Here the scale-up to full production capability is completed. The acceptability of the product for full production is confirmed and final preparations for launch are put in place.
6. *Launch phase.* This phase sees full engagement of the product with the market-place. During this phase confirmation is made that the product meets the quality, cost and customer satisfaction requirements and an assessment is made of market acceptance and field performance.
7. *Maintenance phase.* This phase is used to optimize the revenue from the product and make plans for the end of life of the product based on replacement products expected. Ultimately product withdrawal from the field marks the end of the product life.

The phases are configured as described above largely to facilitate management reviews to control the progress of product delivery. At the end of each phase a set of criteria must be met before the programme can be allowed to enter the next phase. This ensures that the fine balance between performance and customer requirements is kept properly under control throughout the life cycle of the product.

Preconcept Phase

The preconcept phase is the first phase of the product development cycle. It is initiated by the publication of the preconcept requirements in the form of a document that represents the best information about what marketing staff require as a product to meet the customers' requirements. During the phase the business needs and market opportunities are identified, on the one hand, while on the other the capabilities technically are proposed according to technologies available to meet the needs and the technical community's ability to provide a package including architecture that satisfies the requirements.

The phase is used primarily to show a compatible balance between what is required by marketing staff and what can be provided technically. Obviously this requires a clear definition of needs and equally clear proposals for the technical response. It includes the identification of the business needs and the adaptation of these needs by marketing staff to satisfy the market and customer requirements. A technical team needs to be established that can work through the phase to establish architecture and other technical proposals. This team works with marketing staff to establish a set of goals that will meet the needs. Once these have been established the team can work with its technical options to make the necessary decisions to identify the prime system architecture and the combination of technologies, or technology set, that satisfy the business requirements and the programme strategy. A plan to demonstrate technology readiness is prepared and all the programme enablers, including resources, facilities, tools and training, are outlined. In establishing the primary technical approaches the manufacturability and service requirements are considered and fully understood. The whole strategy as defined by this phase of the programme is communicated to the sales and marketing staff to confirm that customers' requirements will be fully met. Finally, a formal review to confirm that all these aspects have been thoroughly thought through and completed is held to give approval to go through to the next phase.

A number of key areas in marketing, engineering, planning and manufacturing are important through this initial phase of the programme. Firstly, there is a need to develop the market and customer requirements. This is primarily the responsibility of the marketing group and should study carefully the customers' priorities and sensitivities and identify the application and feature requirements that will ensure that the product will ultimately meet the customer satisfaction outlook and long term business objectives. This will enable the team to develop realistic goals for the programme and lead to the development of technical options to satisfy these goals.

It is important throughout this phase as the work develops to readdress periodically the marketing and sales responses to the assumptions and preliminary work. If this is left until the end of the phase it invariably turns out that there is further work to be done to achieve closure between marketing and the technical community. Another key area is to establish a business case that is viable and that supports the business objectives of the company.

The technical work should be predicated on the best of what has gone before. This means that each element of the technical work should be subjected to a benchmarking process. This will provide the basis for the boundaries in terms of cost, quality and time to market, and will enable the process of selection of the various technologies to go ahead effectively and unhindered.

It cannot be stressed too much that the goals for the product must be clearly and fully defined. During this phase the goals should be listed at an early stage. As requests and information come in throughout the phase this list should be updated so that at any point in time a distinct list of what is assumed to be in the product goals can be stated.

The strategy for the programme in terms of engineering, manufacturing, service, industrial design, safety and customer support in the field should all be identified during the phase. There will be choices to be made in all of these, which must be made as early as is possible in the programme so that the team members who are dependent on them are in no way unclear about what is happening.

As always, a good and comprehensive plan is an essential enabler to the success of the phase. During the preconcept phase a plan for the next phase, the concept phase, should be developed. This will also lead to the planning for the whole programme which should also be completed at this stage. The high-level programme plan will incorporate the key elements of the programme and will reflect the programme strategy. It will include the number of hardware iterations planned, the level of hard tooling at each of the hardware builds and of course the date planned to launch the product.

The engineering team has a mass of information to collect and process. A search activity related to competitive products, the face-off products and related patents and licences is required to be initiated. As sufficient information is assimilated, the various technology candidates will emerge to meet the product goals. It is often advisable to conduct a technical assessment of the programme fairly early on to establish exactly what further work needs to be done to reach a level of information that will support the right decisions and to repeat this at regular intervals.

A technical feasibility assessment will review the architecture alternatives and the integrated technology set options, and evaluate these against the quality, cost and delivery goals set. To aid the evaluation of the technology set candidates, technical feasibility rigs will be used and these will be aimed at subsequently demonstrating technology readiness, although this will not necessarily be achieved in the preconcept phase.

A design of the product will be started and it is essential that this design adequately reflects the production intent design. If the production intent design is not pursued it will inevitably be found that some of the early work will not be valid and will have to be repeated, thereby extending the overall schedule of the product. The early commitment to production intent in all design work means that some problems and refinements which are associated specifically with the production processes will be addressed early in the programme and will not appear late as surprises that cost money and time to resolve.

It is also useful during this phase to define the full set of activities that will be necessary to demonstrate technology readiness for the product. This plan should outline in particular those technologies of high risk or that have risky manufacturing processes associated with them. Particularly software and electronics, where applicable, should be included.

The process of defining critical parameters should be initiated and combined with the full function diagram incorporating noise factors, response variability, latitude and failure modes. This work may be backed up by using integrated technology rigs to support the data and evaluate further all of these factors.

Combined with the activities in engineering, it is essential that a supporting and compatible manufacturing strategy is also formulated. This obviously has to take into account the attributes of the product such as quantities, life, market geography, cost, quality levels and delivery. Even at this early stage in the programme the manufacturing team should become involved with the engineering team and begin to look at the manufacturing producibility requirements for the product. There may be specific strategic approaches within the manufacturing brief, such as design for assembly, common parts with another product or an environmental requirement such as recyclability. These requirements can begin to be dealt with during the preconcept phase with a combined engineering/manufacturing approach.

At the same time the manufacturing emphasis must lie with producibility requirements, analysing design for any new manufacturing technology needs and focusing on the critical parts, processes and materials that will be required to deliver the product. This may even involve some liaison with suppliers to confirm manufacturability. Overall it is during this phase that the manufacturing engineers should be satisfied that the design approach proposed is consistent with the manufacturing strategy.

Ultimately during the preconcept phase the manufacturing management will be able to provide some estimates of resources, investments and manpower requirements. Preferences for the manufacturing site locations should be evaluated through a comprehensive financial and operational analysis.

Concept Phase

The concept phase is the second phase of the product development cycle. During this phase the readiness of any new technologies being considered for the product is demonstrated and the design for the production intent product is started. An integrated plan is developed which serves to show the way in which the programme will meet its targets not only in terms of schedule but also of cost and reliability. A completed business case showing product viability with the involvement of the marketing and sales operations is established.

The programme at this stage must have a clear definition of how and when it will be launched and this may require dedicated management attention to develop a rational strategy. The strategy will take account of the activities necessary to effect a launch in the business community and the activities here should be matched carefully with the technical activities. This will include such things as sales training, pricing, service and customer documentation and many other post-launch support functions.

Remember that during this phase of the programme the design will be completed and ideally frozen. It is therefore essential that the marketing strategy be completed and clearly defined. All marketing aspects such as target markets, product positioning, pricing, servicing and customer support strategies must be understood and finalized. A market forecast to provide essential data to manufacturing should be evaluated. Other business goals to enable the development of engineering and manufacturing targets should be quantified. These data will enable the full business case to be prepared and evaluated.

For the concept phase the full development engineering and manufacturing team will be set up and with the product goals finalized by marketing staff the technical task of product development can commence. The programme plan will inevitably require a significant update, not only to reflect any changes brought about during the previous phase but to encompass the more complete understanding of the overall strategy which will now have developed. Part of this plan should be the setting of targets for the team to work to in order to meet the requirements of the programme. This enables a clear understanding for the team in terms of boundaries within which the programme must be constrained.

During the concept phase it is of course essential that the design team looks forward and develops a plan for the next phase. The plan should include the objectives for the phase, the baseline criteria for the production intent design, the phase exit criteria and of course the individual activities of the phase which are required to meet the objectives. An integral part of this plan is the actual scheduling of the activities and the resources and training needs that are required. In fact, it is a good idea to prepare a list of all the enablers for the phase at this stage.

The recommended review process is described in Sec. 4.6. In order to establish the programme's readiness to move to the next phase, a review of the programme status is required. This should follow the lines described in Sec. 4.6, and should address the following key areas:

Technology
Business and product planning
Finance
Engineering (reliability, design, software, performance)
System engineering
Manufacturing
Marketing
Product appearance and operability
Programme plans and resources

During the concept phase the competitive situation can be assessed and the response to this situation decided upon. This means deciding just how the product under development will be positioned within the existing product scenario and how this will challenge or divert the competition. One useful process to evaluate the competitive picture is to model the product as a 'best of breed'. This means looking at the whole of the competitive picture, choosing from each competitive product the best attribute (or attributes) and then building a model that brings all of these attributes together into one product. For example, one competitive product may have a particular feature which sets it apart from the other products in terms of customer appeal. Another may have a low cost attribute. Another may exhibit high reliability. A 'best of breed' model would bring together all of these 'best' attributes into one product. Obviously such a model rarely becomes a product, but it enables the team to decide which of the excellent features or attributes should be pursued. This exercise should be completed and documented in the concept phase.

The programme specifications and system requirements are issued so that the design plans for the production intent design can be prepared. A systems operation description, which describes the functional operation of the product, is prepared. This is an essential document to bring together the individual elements of the systems operation and enables development of the software to support this. Similarly, a plan that brings together and addresses the interactions of the various engineering elements of the programme should also be prepared. This pays particular attention to the systems elements of the product development.

An integrated test plan will identify all the test procedures and methods to be used to validate the system and subsystem performance. This plan establishes the criteria and measurements which will validate the readiness of the product to move to the next phase. A software development plan is also required.

Technology readiness is required to be demonstrated during this phase. This must address:

The identification of critical parameters
The identification of failure modes
The demonstration of systems latitude
The demonstration of the production intent design hardware
An understanding of manufacturability

During the phase the production intent design is started. This will comprehend the system and subsystem requirements and represent the decisions made on the 'best of breed' model. As part of the development process, manufacturing, service and other requirements are incorporated and optimized. Designs are optimized and evaluated against the specifications using rigs and fixtures.

A design analysis is completed so that the following assessments can be made:

Assess how well the design approach meets customer requirements.
Identify risks and opportunities.
Assess how well the programme meets its quality, cost and delivery criteria. Targets for cost, quality and delivery are defined.

It is very important that during the concept phase an adequate manufacturing support team is in place. The manufacturing support team is responsible for defining manufacturing strategy for the product, developing the manufacturing plan and implementing the necessary manufacturing processes required to deliver the product. This includes:

Programme planning assumptions
The definition of roles and responsibilities
The definition of the programme schedule
Site selection for production
A plan for materials acquisition
A quality management plan
A capital plan
A manpower plan
Configuration control of the product

Design Phase

The design phase is the third phase of the product development cycle. During this phase, the production intent design is completed, and this will include the full feature and function set of the proposed product. Programme specifications are finalized with resolution on all the specification issues. The production intent design is built and tested against the programme criteria to demonstrate design stability and control of the configuration. A programme launch strategy is developed and the sales units are prepared for the launch of the product by becoming involved in the programme during this phase.

In this phase of the product development cycle, attention is centred on the launching of the product, what the criteria will be and how and when the product will be launched. This approach is necessary simply because there is an activity that is parallel to the design activity which has to start early in order to effectively launch the product. This activity will be dependent upon exactly what the programme can deliver and when. Consideration is therefore given to the criteria for launching the product so that such things as sales documentation and training manuals can start to be prepared. Support from the sales unit will be required and so confirmation of how the new product will perform and when and in what quantities it will be available is made based on agreed sales support. Once again it is prudent to update and review the business case associated with the product to take account of any technical or marketing changes that may have occurred.

The design phase plan prepared during the concept phase is now implemented. This plan will identify any lessons learned from previous products or programmes, assign clear responsibilities for all the team, and address all the known technical problems currently listed.

In the area of engineering, the specifications at this stage are frozen and can only be changed by a formal and controlled process. Product appearance is finalized. The production intent design is completed and the first iteration of the hardware for testing during the phase is built. This includes any software that may be necessary as part of the first hardware build. Ideally the design will become stabilized quickly during this phase as changes from now on will become costly both in financial and schedule terms.

The capability to stabilize the design adequately will depend on how well the previous phases have been implemented. This will depend to a large extent on how well the enabling techniques described throughout this book have been achieved.

The hardware is subjected to a systems test and its performance compared with the programme targets and customer requirements. Problem management processes are used to provide rapid solutions to the problems identified during the systems test. Again many of the techniques described throughout this book can be employed here.

Finally an assessment against the programme's predicted capability to meet cost, performance and delivery expectations is carried out and used as part of the assessment study to decide whether the programme meets the phase exit criteria.

Demonstration Phase

The demonstration phase is the fourth phase of the product development cycle. The central objective of this phase is to demonstrate the stability of the production intent design and the readiness of the production processes to deliver it. This is accomplished by building a pilot or preproduction model and testing this against the criteria for launching the product.

During the phase there will be a plan for all the necessary hardware, spares, accessories, etc., to support launch. This is provided by the sales/operating units of the company. A launch strategy is also required which covers the following details with respect to launching the product:

- The strategy for launching the product including distribution strategy, quantity profile and geographical limits of the launch process
- Customer service training and documentation
- Spares strategy
- Pricing strategy
- A plan for getting early customer feedback on the product performance
- Advertising strategy

At the subsystem and system level the engineering team should be focusing on the identification of performance-related problems and their solutions. They should validate the functionality of the product, the design latitudes and the acceptable ranges of all the critical parameters. In association with the manufacturing team there will be a combined effort to develop and stabilize the production assembly processes and to validate the manufacturing assembly process in a production environment. This is a joint venture between engineering and manufacturing and the success of it will depend very largely on the cooperation between the two. It requires the facilitation of 'concurrent engineering' practices to mature the design and develop the manufacturing tools in parallel with the build of the hardware.

The percentage of hard tooled parts for this build of the hardware is important. The higher the percentage of hard tooled parts at this stage, the more representative will be the results from the build and test stages. Coupled with this, the minimization of variability in the parts is a key enabler for the success of this phase.

Once the pilot production hardware is built and debugged, a system test is performed aimed at understanding the effects of the production processes on the functioning of the product and evaluating any new design changes that have been incorporated. When the results of these tests are evaluated an assessment is made as to whether the current level of design and manufacturing processes are on target to meet the product mature goals. If this is not confirmed, then further design and manufacturing improvements may need to be made before proceeding.

Production Phase

The fifth phase of the product development cycle sees the full scale-up of production for the product and testing to verify that the manufacturing processes in this form are meeting requirements. During the phase the prime responsibility for the product programme remains with the chief engineer and the product delivery team comprising engineering, manufacturing, marketing and service personnel, together with other support personnel as appropriate. Once again, the plan for launching the product is brought up to date by encompassing any changes which may have occurred in marketing, quantities, service strategy, etc.

One of the key activities during this phase is to gain management approval for the programme to go from the production phase to the launch phase. This is a particularly critical review due to the very large amounts of money that will be committed and the short time involved in this expenditure due to the very high rates of production and the costs that go with it. The review will take into account:

The latest business case
The engineering reliability, design stability and performance
The manufacturing quality
The marketing, customer support and service readiness
Customer acceptance feedback

This will be assessed by significant testing of the production products on an in-house system test, together with customer acceptance testing with selected customers to validate the level of customer satisfaction with the product. Of course this may involve engineering and manufacturing in some late problem solving which may initiate changes to the design. This should obviously be kept to a minimum but will feature heavily in the assessment for launch readiness of the product.

All the necessary preparations for launch are put in place and confirmed during this phase which of course involves not only manufacturing and marketing departments who are supplying the product but also those who are receiving the product, such as sales and servicing.

Launch Phase

The launch phase is the sixth phase of the product development cycle. During this phase the product is introduced to the end user through the implementation of the launch plan for the programme. The phase confirms the performance of the product in the field and takes a reading of the customer acceptance of the product compared with other products, both own and competitive.

Once approval has been given to launch the product the launch plan is brought into operation. In each sales area the data connected with the launch plan, such as the number of units placed versus the plan, are monitored, together with feedback on the performance of the early units. At this stage any problems identified in the early stages of the launch phase must be dealt with quickly and effectively so that products being produced (now in large quantities) can have the problem successfully eradicated before it gets into the field.

Customer satisfaction monitoring should be maintained throughout the phase so that all aspects of the customer satisfaction shortfalls can be addressed quickly or at least recorded for attention in designing the next product. In this respect, it is a good strategy to conduct a 'lessons learned' study during this phase. This may promote some enhancements to the existing product which could be a significant enabler in raising customer satisfaction levels. This in turn may trigger the design of a variant product which will also extend the life of the original product at the same time as improving its performance. At the very least, a lessons learned strategy will enhance the quality of the next product and the delivery of it.

Senior management approval is required in this phase to transfer the product into a maintenance phase. Such approval reflects successful completion of all the objectives established for the launch phase.

Maintenance Phase

The maintenance phase is the seventh and final phase of the product development cycle. During this phase the production of the product continues to meet the full demands of the customer and it is the phase where the peak of the revenue and profits for the product is taken. During this phase, the end-of-life strategy for the product is established.

In this phase the chief engineer and the product delivery team relinquish the design authority for the product, but are retained in a support role. Design responsibility is transferred to the manufacturing function. An end-of-life strategy is established. This includes:

Schedule for the process to wind down production and bring up the programme for the replacement product
Customer communication strategy
Inventory and spares management and control
Pricing strategy

The lessons learned from the programme are updated and together with any new requirements from marketing a strategy for replacing the product is developed. This strategy may use an enhancement to the existing product or form the basis for a completely new programme. Some products may lend themselves to refurbishment, which may also play a part in the product replacement strategy.

Finally, there will always be a few products left in the field that require maintenance and spares. In order to support these a strategy for maintaining a minimum number of spares will need to be put in place.

4.2 ROOT CAUSE ANALYSIS

Even with an efficient design process in place, competition is always snapping at the heels of even the best companies and there is an overriding need to put in place a process of continuous improvement. How can this be done? Where does one start to look at what is currently the method of operation and where can improvements be made? Learning from the mistakes of the past is a well-tried method of making improvements over what went before. However, a process that has been internalized into the design process itself so that there is an automatic discipline for making continuous improvement is even better.

The following method is in many ways generic and can be applied to any process that is defined, not just the design process. It is a method that I have seen work successfully over the years to bring about change to eliminate errors. It is based on the principle that, by using data to measure the effectiveness of the existing process, the process can be changed to improve the effectiveness further.

The process uses the following steps:

1. Decide on a factor that can be used to measure effectiveness (or lack of it).
2. Define the factor in measurable terms.

3. Define the current process and identify the points where progress is checked.
4. Identify examples of where effectiveness is inadequate.
5. Identify which checkpoint failed to expose the shortfall.
6. Modify the process to rectify it.

As an example the design process is considered to be measurable by monitoring the number of design changes per thousand parts per month. High levels of change have a significant impact on the progress of design against its planned schedule, the more changes that are required causing a slow-down in the design itself. High levels of change also incur extra cost to the development of a product. Design changes therefore seem to be a suitable factor with which to measure the effectiveness of design progress.

The design process being used can be clearly laid out in full detail in the form of a flow chart, paying particular attention to the various checkpoints that are used as part of the process or as gates to stop the programme moving on to the next phase unless it has reached the necessary criteria (Fig. 4.2). The next stage is to collect data on the full set of design changes that are on record and identify the root causes of the need for the change. This identifies the basic problem that caused the change and in most cases it is easy to estimate when this problem should have been identified in the normal course of events. A matrix can then be used to plot each of the problems to depict (a) at what stage the problem was found to require the change and (b) where the problem should have been found in the ideal world. As can be seen, the matrix clearly shows the relative lateness of identifying problems against an ideal case (Fig. 4.3).

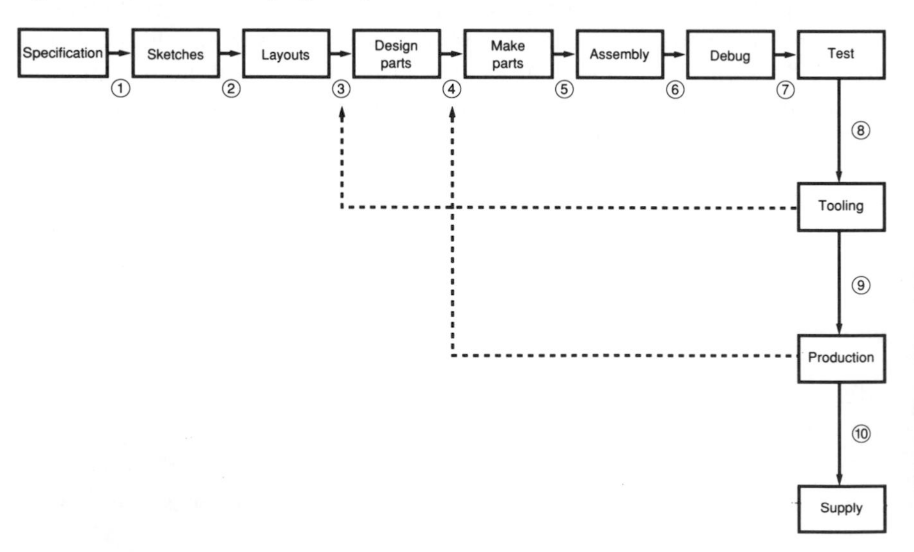

Figure 4.2 Design process flow chart.

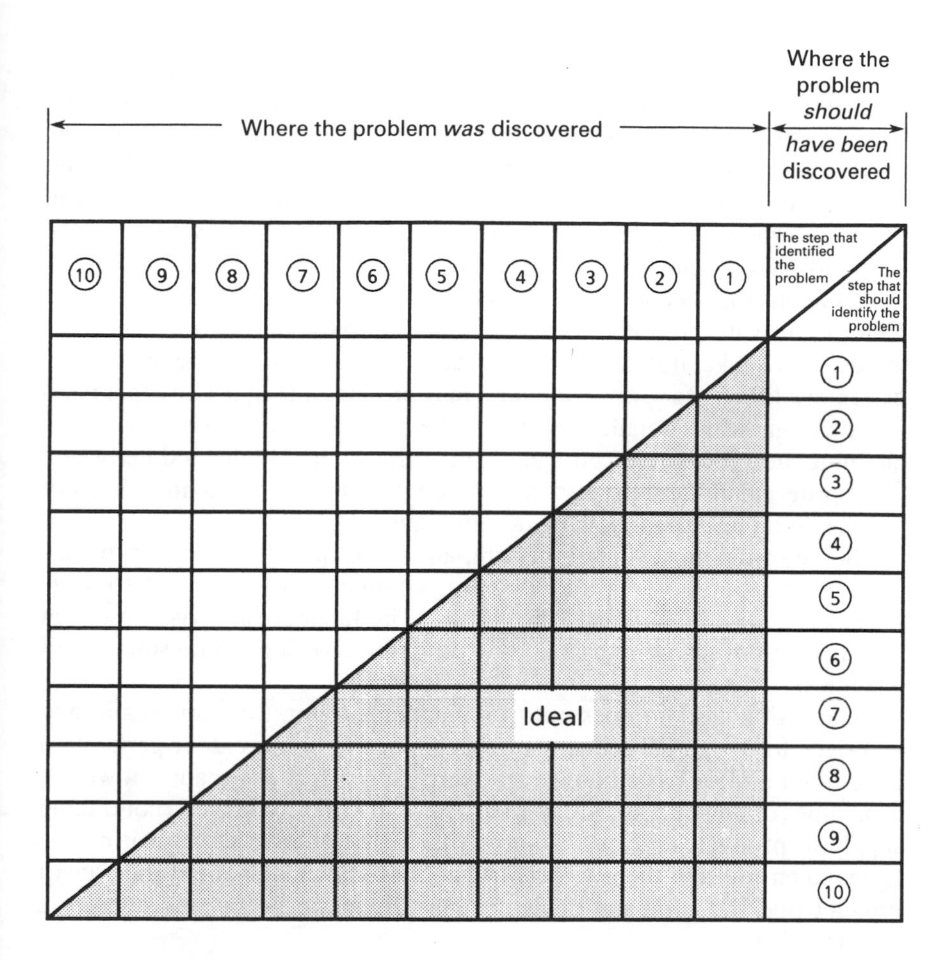

Figure 4.3 Problem identification matrix.

The information on problems found by this method can be used to highlight which checkpoint in the design process failed to detect the problem at the right time. Once this is known, and once the reasons for this are known, it is a simple logical step to identify what should be changed in the process to prevent this happening in the future. Thus the process can be continually updated to take account of shortfalls of the past and eliminate them for the future.

Once this method is understood by the engineering team the work involved in implementing it is minimized, as the process described above can be adopted as a part of the normal problem solving process for the programme.

4.3 CAPTURING WISDOM

A study in the United States addressed the problems of large corporations compared with their Japanese counterparts and specifically investigated the reasons why design teams appeared to be unable to maintain the schedules of programmes against the plans. The results of this survey led to a study of the total product development process, but one of the key factors that emerged was that much of the time wasted was due to mistakes made by the design teams that needed to be corrected later, often at a time that caused major upheavals, not only to the schedule but to the design itself. In other words, a mistake not found in the early stages of the design may be extremely difficult to resolve later and may produce an impact on other parts of the design. More detailed studies of the mistakes revealed an even more alarming fact. A high percentage of the mistakes had happened before on other programmes and the lessons learned from these programmes had not been passed on to the following design teams.

Some of the most successful companies and those producing products carrying a reputation for quality and reliability are perceived to have engineering wisdom built in to the culture. The Japanese are past masters at learning lessons from the past and feeding this knowledge into subsequent activities.

Capturing wisdom does not 'just happen'. Using the same team over and over to design subsequent products does to some extent generate a certain amount of basic knowledge which stays with the team. However, companies often need to overlap their teams in order to deliver a continuing supply of products and this requires that a discipline is set up within the teams to ensure that there is a central pool of knowledge about the type of product that the company deals in. This discipline needs to recognize a number of facts:

1. The designer is the expert in a particular field within design.
2. Whatever discipline is introduced, it must take input from the experts and make it available to other experts.
3. It must be self-perpetuating.
4. It must be evergreen (always up-to-date and being improved).
5. Ideally it should be capable of being used as an expert system.

There is an interesting analogy to this proposal. The designer is the expert in a particular field of design in the same way that a pilot is an expert in flying an aeroplane. We give the designer the tools in the form of methods and procedures to do a job. The designer is trained in the skills of designing and the technical information we ask him or her to use is viable. When we board an aircraft as a passenger, we assume the pilot is trained and has the technology to fly. In spite of all this we recognize that even a pilot can be subject to human error. Therefore, before a pilot takes off, there is a check-list to run through to confirm that all actions for which the pilot has been trained are properly carried out and that all of the essential points that will make flying the aircraft safe are not forgotten.

Using check-lists in design has been shown to be an effective way of improving quality and minimizing mistakes.

Design Quality Guide

The design quality guide is equivalent to the pilot's check-list. It is accepted that the designer is fully trained and capable of good design, but it is also recognized that the designer may make a mistake. The design quality guide is intended to eliminate this as far as possible.

What the design quality guide contains is a series of questions aimed at alerting the designer to points, not necessarily connected with the basic technology, that are known to be important for a good embodiment of the design. The knowledge comes from the experience of other designers and engineers and is often the result of solving a problem on a previous programme. The process of finding a solution will have been exhaustive and the result is likely to be robust, especially as the programme will by now most likely be proven in the field.

A typical guide is depicted in Fig. 4.4. It consists of a series of questions pertinent to the area of design, the questions being generated initially from previous experience of problems and their solutions. For example, it is known that using a sintered bearing with a stainless steel shaft demands that the lubricant which is impregnated into the bearing contains a particular small percentage of molybdenum disulphide, otherwise the bearing will very rapidly begin to seize and subsequently fail. It is the easiest thing in the world for a designer to call up standard impregnated bearings, either if he has not experienced this at first hand or if no one has informed him of the fact. A suitable question in the design quality guide might be, 'Have all sintered bearings been specified with the correct lubricant where used with stainless steel shafts?'

The questions will have been updated and added to by designers using them and adding their own experience to the total list. As the guide becomes well used some of the questions will become redundant as the users will have internalized the information in their normal design process. When this happens the questions can be deleted from the list, and as new problems are resolved new questions will be added to the list. In this way the guide maintains its effectiveness and remains 'evergreen'.

Ideally, in a design domain that is computer aided the guide with its questions should automatically prompt the designer while proceeding through the relevant section of the design. Eventually the guide may become part of an expert system that supports the specific area of the design and ensures total quality of each application.

Paper transport rolls and guides

Critical parameter information

Have all critical parameters been listed? Yes____ No____ Contact person____________ Phone No.____________ Mail code________

Have nominal values been assigned? Yes____ No____ Have tolerances been set for each of the nominals? Yes____ No____

Have failure modes been identified for conditions outside the tolerance window? Yes____ No____

Has the analysis of the manufacturing process/tolerance build-up been completed for each critical parameter? Yes____ No____

Critical parameter audit P1 P2 B0 B1 B2

Design guide

	Yes	No
Architecture		
1. Is there an adequate number of rolls to transport smallest size paper?	____	____
2. Has the minimum radius in the paper path been determined by using the recommended limitations of radius for different paper weights?	____	____
3. Are roll segments positioned to deal with paper width set?	____	____
4. Has distance between registration and transfer been determined?	____	____
5. Have sensor positions relative to nip rolls been determined?		
Is wait station switch before the nip rolls?	____	____
Is hold station switch after the nip rolls?	____	____
Is position on monitor in correct position relative to jam clearance strategy?	____	____
Guide and roll configuration		
6. Have you referred to the standard drawing regarding nip roll cut-outs?	____	____

Design Notes

Figure 4.4 An example of a design quality guide.

The design quality guide can enhance the chief engineer's or project leader's view of progress. By assessing the compliance level of a particular drawing to the check-list requirements, the manager can gain a good indication of how much progress is being made. Also, using it as a monitoring tool, together with other monitoring tools, the design quality guide gives the manager the capability to make good decisions on whether to release drawings for manufacture or not.

In addition to providing a means for reducing design errors and enhancing quality by drawing on the experience of previous programmes, the design quality guide also serves another purpose. As the company grows in maturity, the wisdom gained by successive design activities also grows, and it is essential that this wisdom is captured for the future. As expert systems become more commonplace, there is a need to create a cycle of knowledge transfer so that any knowledge base that is in place can continue to grow and change as technologies in all areas of engineering change. The people who have the knowledge and are the experts are the designers and engineers themselves, and therefore it is essential that a system exists to enable a continuous flow and update of their knowledge into expert systems. The design quality guide is one way in which this can occur.

4.4 PEER REVIEWS

How do we assess quality, cost and the ability to meet the schedule during the design process? A number of methods and procedures for keeping track of design progress have been successfully implemented in the past and each of them brings some additional knowledge and confidence to the status. Many are simple check-lists which enable the designers to double-check against their own expertise and experience that the quality of their design are acceptable. A popular method used with success is to use other designers and engineers who are not directly involved in the design itself to review the design periodically in an informal way. Often these informal reviews are called 'board reviews', dating back to when most design work was completed on a conventional drawing board. The use of computer-aided design does not prevent us from using the same style of informal review. The reviews are usually organized on a regular basis to review in general terms the progress of the design work. Often, however, when a particular problem is uncovered the same style of review can be carried out to take a more specific look at the problem area.

The review takes the form of a number of designers, engineers and technical managers gathering around the workstation (or drawing board) so that the designer can get the benefit of their experience, who although not directly involved in this design have had experience of design or engineering on similar systems. The basic premise for the exercise is that more heads are better than one and, as such, mistakes can be avoided or new ideas introduced by the very fact that an outside look at the design sees it in a different perspective. Additionally, designers can often develop a very narrow view of their work and either fail to appreciate some of the peripheral points of the design work, such as servicing or tooling pitfalls, or have difficulty solving some of the problems that are faced in the design.

Some of the benefits of board reviews are listed here:

1. The designer receives the benefit of additional design experience from peers which may be synergistic with the design.
2. Regular board reviews in all programmes enable ideas to be automatically shared between design teams without the need to put in place a formal process to do this.
3. Lessons learned from other programmes quickly eliminate mistakes and/or supply the answers to problems that might otherwise take a major effort to resolve. This happens earlier rather than later which is a double bonus for the programme.
4. Cost is generally reduced because changes to the hardware are reduced and the overall time and resources to solve problems and provide solutions to the hardware is significantly curtailed.
5. Schedule is also often a beneficiary of this as a result of less changes and earlier identification of problems.

4.5 MONITORING AND TRACKING DESIGN

During the intensive period when the designing is actually going on, there has always been difficulty in determining just how well the process is progressing, whether one part of the design is proceeding at a greater rate than another and whether some change in resources or plan is required to enable the design team to meet the targets. Progress is usually measured by reporting the number of details issued. Hence the first indication that there are schedule problems is when the rate of production of detail drawings falls short of the planned rate, a time much too late to take corrective action. Quite often the lead designer is responsible for reporting progress to the manager. Traditionally, the lead designer will be asked whether the work of the team is progressing on track by treating it on an individual drawing basis. Usually the designer will have a feel for how much progress is being made and will estimate (more often than not) some slippage. It is not unusual to find that these estimates are wildly out from reality. My experience has been that often a major iteration of design can be as much as six weeks late to the plan with no apparent problems to overcome. On one programme the team was asked to estimate the drawing completion date for each of the drawings being prepared. They were asked to give their most pessimistic dates and then asked to make stringent efforts to keep to this plan and highlight any shortfalls expected as early as possible. In the event the drawing iteration was the traditional six weeks late as before! So there is an urgent need to find a process that will enable the designer to (a) plan the work realistically and (b) have the ability to monitor progress so that early indications of failure to meet the plan can be identified and resource or plan changes introduced to compensate for this.

The process proposed here depends on breaking the overall package of work into subtasks which can then be built into the overall programme schedule. The method is flexible in that it allows the work to proceed in any order and can enable the lead designer to redistribute the work within the team to compensate for any shortfall in one particular area that may have been caused by an unforeseen problem. It requires the knowledge of an experienced designer to be able to break down the total workload into modules that are measured in quantifiable units relating to their complexity and hence the length of time required to accomplish them. The method has the following attributes:

Simple to use
Quick to update
Allows work order to change
Allows work content to change
Allows staffing levels to increase or decrease
Allows the effcct of schedule changes to be forecast
Allows the tracking of tasks that cannot be accurately defined in detail but have to fit within a given schedule

Design Tracking Procedure

1. Identify the drawing tasks to be completed by describing them in the smallest unit modules possible. This of course depends on the ability of the designer to assess the work in terms of complexity, likely difficulty and time to complete. Ideally the number of task modules should exceed the number of designers working on them by a factor of ten.
2. Select the shortest task of all and give this a value of 1, representing the time that will be required to complete it. This will be regarded as the base task.
3. Compare each of the tasks in turn with the task rated at value 1 and give it a numerical rating based on the length of time expected for completion compared with the base task. For example, a task that will take three times as long as the base task should be given a rating of 3.
4. Sum the task ratings for the complete job.
5. Make an arbitrary allocation of tasks to each of the designers in the team.
6. Sum the ratings of the set of tasks allocated to each of the designers. For designers of equal ability, the sum of the task ratings should be roughly

equal. For designers of lesser ability estimate the ratio of time taken to do a standard job versus that of the more experienced designer and try to arrange the total task ratings in the same ratio. If a designer is expected to work part-time then the total task ratings should be in the correct proportion.

7. Plot a graph as shown using a base of time and a vertical axis of task units. Identify the time at which the work is required to be completed and show 100 per cent task units complete. The curve can be plotted either as a straight line or using a more realistic 'S' curve (Fig. 4.5). Similar graphs can be plotted to define both team and individual activities.
8. As the design progresses the completion or estimated part completion of each module of the work can be depicted on the curves and compared with the plan line.
9. Any shortfall in an individual's expected progress will be seen very quickly. It may be possible to reallocate some of the work that was allocated to this designer to someone else on the team. Any team shortfall in progress will also be evident quickly and this will alert the design team to take some positive action to resolve the situation.

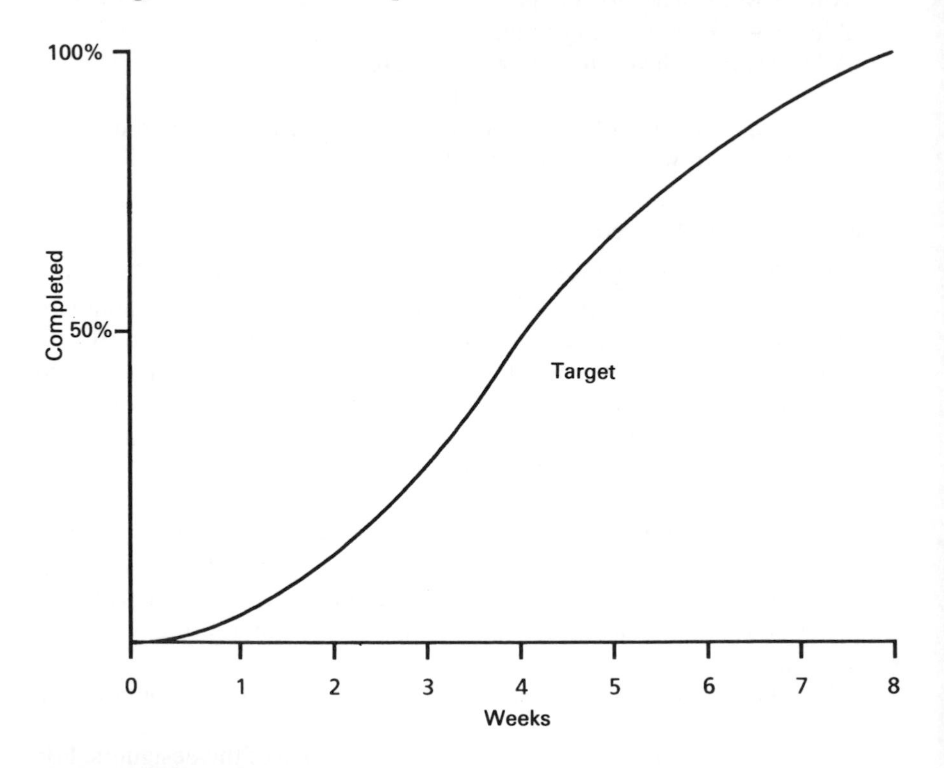

Figure 4.5 Work progress chart.

Planning					Monitoring at date . . .						
Design tasks	Task rating	Designer task allocation			Progress (%)	Completed work			Work to be completed		
		A	B	C		A	B	C	A	B	C
Paper feeder	4	4			50	2			2		
Separator	4			4	75			3			1
Transport	5		3	2	40		2			1	2
Elevator	2	2			0				2		
Drives elevator	3			3	0						3
Drives separator	4	4			50	2			2		
Drives feeder	2	2			100	2					
Chassis	3		3		0					3	
Harness route	1		1		100		1				
Doors and interlocks	5		2	3	20			1		2	2
Top guide assembly	3	3			100	3					
Lower guide assembly	5		5		80		4			1	
Retard roll assembly	1	1			0				1		
Transport sensors	3			3	66			2			1
Total	**45**	**16**	**14**	**15**	**49**	**9**	**7**	**6**	**7**	**7**	**9**
Column 1	Column 2	Column 3	Column 4	Column 5	Column 6	Column 7	Column 8	Column 9	Column 10	Column 11	Column 12

Figure 4.6 Design progress chart.

Figure 4.6 Design progress chart.

Example of the Process

1. List the tasks on a chart as shown in Fig. 4.6.
2. 'Harness route' is selected as the standard being the shortest defined task.
3. 'Paper feeder' is estimated to need about four times more design resource than 'harness route' and is therefore given a rating of 4. Enter this in column 2.
4. Compile the rest of column 2 by repeating this for each of the tasks. The total adds up to 45 task rating units.
5. Assuming that three designers are available, proportioning the work out equally would allocate 15 units to each designer. Of course some designers may be capable of working faster than others so the lead designer may allocate the tasks disproportionately.
6. Allocate the tasks and show this in columns 3, 4 and 5. This completes the planning stage of the process, the other columns of the table relating to monitoring progress.
7. At any point in time, the lead designer can track the work by estimating how much each designer has completed on the allotted tasks as a percentage. The amount completed for each task is entered in column 6.
8. Columns 7, 8 and 9 show the work still completed individually by each designer. Columns 10, 11 and 12 show the work still to be done. By totalling this for each designer the lead designer can reassess the capability of each of the designers to do the job. This may involve some reallocation of tasks.
9. Plot a graph (Fig. 4.7) to show the overall progress of the group on a weekly basis.
10. Using this graph (and if necessary similar individual graphs for each designer), progress can be monitored regularly to enable early changes to be made if necessary.

The process is simple but it serves to ensure that a good control is maintained on the progress of the work so that any changes required in either priority, work allocation or size of task can be implemented immediately and a revised plan put into operation.

4.6 THE REVIEW PROCESS

From time to time there is a need during the design process to conduct a formal review of the status. This is usually carried out as the design reaches the end of a specific phase and the information and understanding gained from the review are used to make decisions on whether to go on to the next phase or whether to make some changes in the programme strategy.

Design teams usually dislike the ideas of reviews because it seems that

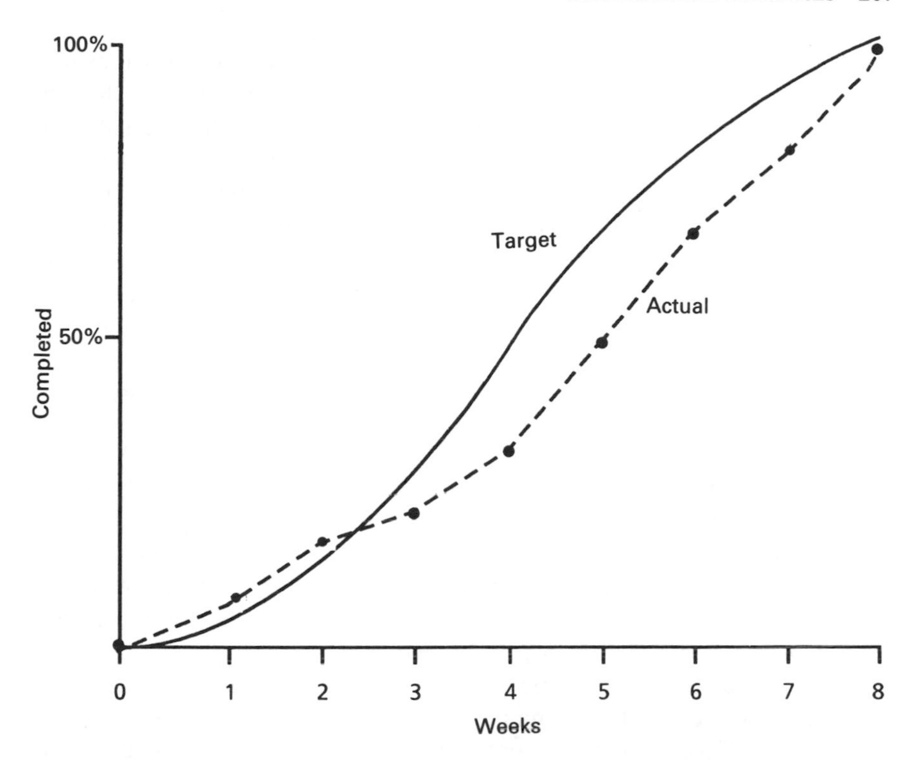

Figure 4.7 Progress against target.

they are required to do a lot of work in preparation and it interferes with the work that they know they have to do. I believe, however, that reviews are not only necessary but can also be useful in enhancing the status and direction of any programme. If they are handled in the proper way they can contribute to the overall well-being of the programme and not be regarded as too much of a chore.

In general reviews are useful for the following reasons:

1. They give the team official time to look at their progress from a fresh viewpoint.
2. They enable the status to be documented.
3. They can achieve their main goal to confirm that the programme is on the right track or recommend redirection.

Like all things, there is a good way and a bad way to conduct a review. I have experienced both, but I will set out what I believe to be the best process. Firstly, any review must start with a self-assessment of the programme. The best people to define the status, the strengths and the weaknesses are the people who have done the work. It is a fallacy to believe that an outsider can come into a programme, collect information for a week and then make a judgement on status. If this is attempted the result will usually be that the assessor reads back to the team the very information that the team has divulged and although this can have the same result it is frustrating for the team to hear their own judgements from the mouths of others.

It is essential to allow an adequate amount of time to unearth the information and sift through it. Any attempt to short-cut the process will result in an inadequate assessment of the situation. It is also essential to have an outside independent assessment team or person to conduct the review. This gives the opportunity for all the data to be considered from an independent viewpoint which is invaluable to judging what the true status is.

A typical review process that I have found to be most successful is as follows:

1. The programme team prepares information on status, progress against plan, problems, issues and actions to resolve issues. This preparation is documented and shared with the other members of the team.
2. The team makes a self-assessment as to how they are progressing and ideally measures this using some systematic process previously laid down (see Sec. 3.14, Problem Management).
3. The concurred team position is then documented as pre-reading and sent to the selected assessment team. This team reads the programme team's self-assessment and notes any questions, comments, etc., that they may have.
4. At a prearranged meeting the programme team present their findings to the assessment team, reiterating their self-assessment and answering and clarifying any questions made by the assessment team.
5. Further clarification is then achieved by holding individual meetings between the programme team members and the assessment team members. The assessment team may have their responsibilities divided up and matched with those of the programme team.
6. The assessment team then meets to discuss together their assessment of the programme team's findings. Key points are discussed and if necessary the programme team representatives are once again consulted for clarification. The assessment team then organizes the main issues and problems into a list and makes recommendations as to how these should be resolved. They then summarize the results of the review and recommend in general terms whether the programme should continue on its present course or be redirected.
7. The findings of the assessment team are then presented to the programme team and the programme team is given the opportunity to make a response. Obviously any major disagreements have to be resolved and the primary objective is to reach a consensus between the two teams.

8. Following consensus and, if necessary, the concurred position is transmitted to the authority who can confirm continuing or changing direction of the programme.

CONCLUSIONS

Most of our industries which are involved in designing have been searching and are continuing to search for something that will give them the competitive edge and a more effective design process. In particular, there is a desperate search for a solution to shorten the design and development cycle. Indeed, there is a growing feeling that unless many of our industries improve their performance in these areas they will die in the face of stronger competition. Many techniques have been described in this book and all of these will contribute to a more effective process. There is no single method or process described here that can do it alone, but a combination of these techniques and good management can make a significant improvement in the way products are brought to market.

How should it be done and what are the pitfalls to be avoided? Firstly, nothing will change unless the highest levels of management are committed to make the necessary changes. This point cannot be emphasized too strongly. Secondly, there is an investment to be made both in terms of time and money and most of this will be expended in training. The level of commitment of an organization to both these points will decide the level of success achievable.

Once committed to making changes on these terms it is essential to understand that implementation is 90 per cent of the battle and implementation can only be effective if again there is commitment at all levels to make the changes. It will take time and will initially slow things down to train professional people in these additional skills. Ideally, we need to train them in these skills as a part of our technical education, and there are growing feelings that our universities should include 'design' as a larger proportion of their engineering courses.

A greater emphasis on understanding function at the design phase will enable time and cost to be saved in the overall process. Techniques to define more accurately the requirements of the customer will ensure that a product meets those requirements more effectively. Other techniques, like FMEA, will help to eliminate those failures that give products and companies a bad reputation for reliability. Many of the processes will bring engineering and manufacturing much closer together and as a result will help to reduce time-to-market. Better management processes will enable decisions on investments in new products to be more effective and will enable risk management to become more of a science than an art, so increasing competitiveness.

Using these improved methods, coupled with new technologies like CAD and a more effective management style, will enable industries to compete with the best in the world and deliver quality through design.

SUMMARY OF TECHNIQUES

THE COMPOSITION OF DESIGN

Form
Fit
Function

THE PHASES OF DESIGN

Define the requirements of the customer
    Use quality function deployment
Generate ideas by innovation
    Use brainstorming and selection processes
Understand the function of the design
    Use FAST and function diagrams
Translate the design into drawings or a design file
Transfer the information into manufacturing
    Use simultaneous engineering

THE ARCHITECTURAL PROCESS

Five phases to the process:

Definition
    Outputs and boundary definition established
    Architecture core team formed

Development focus and scope defined
Process support groups identified
Assumptions established
Competitive benchmark strategy established

Concept generation
Functions identified
Concepts brainstormed
Concepts and functions integrated
Study papers identified and written

Concept reduction
Functional requirements reviewed and satisfied
Product specifications reviewed and satisfied
Concepts refined
Concepts eliminated
Candidates selected

Concept evaluation
Evaluation methods selected
Business and technical risks assessed
All relevant QCDs assessed

Output
Conclusions reached
Questions identified
Recommendations made

CONCEPT GENERATION CHECK-LIST

Product theme
Product specifications
Critical design issues
Competitive benchmarking
Relevant standards
Establish minimum level of detail for each concept
Preliminary evaluation criteria

EVALUATION CRITERIA CHECK-LIST

Relevant product specifications
Relevant architectural standards
Technical risk
Business risk

Concept advantages and disadvantages
Effect on other parts of the design
All factors affecting cost, quality or schedule

THE PUGH PROCESS

Use an icon to distinguish the concept
Define a standard that is well known for comparison with other ideas
List the judging criteria
Score with +, − or S
Add up all the +, − and S values
Assess the results
Apply a test of reasonableness

THE COMBINEX METHOD

Define the selection criteria
Apply weighting to the selection criteria
Develop utility curves
Assign points to each alternative
Apply weighting to the points
Make a choice
Apply a test of reasonableness

TECHNOLOGY READINESS

The five criteria for technology readiness:
For all systems and subsystems
- Critical parameters defined
- Failure modes defined
- Latitude defined
- Manufacturability assessed and found acceptable
- System hardware demonstrated

FAST DIAGRAMS

Functional analysis system technique
Function boxes linked by 'how' and 'why' relationships
Ask 'Why does a function exist?'
Ask 'How does a function occur?'

Functions defined by noun–verb phrases
Construct the diagram with 'hows' to the right of the relevant function and 'whys' to the left
Draw boundaries around the functions that represent the focal area of the design study
Draw the critical functional path through the system
Look for repetition of functions

THE FUNCTION DIAGRAM (see Fig. 3.13)

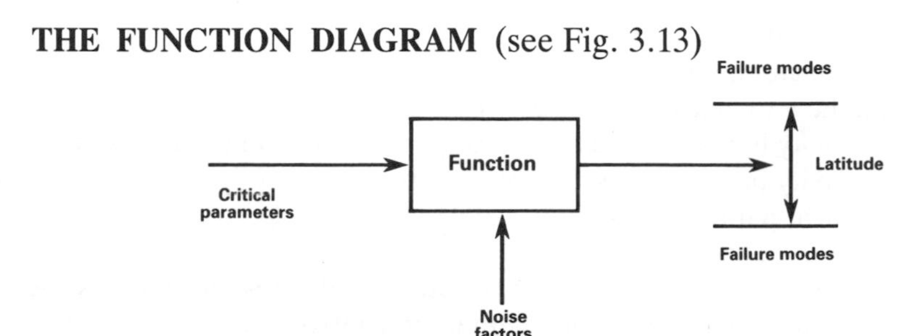

The function must be described in a noun–verb combination. The inputs and outputs to and from the function must be described. They are:

Critical Parameters

These are the parameters that are within the control of the designer and drive the function.

Noise Factors

These are the parameters that have a major effect on the function but are outside the control of the designer. Often these are environmental effects.

Responses

The response of a function is the output from the function. It is what we want the function to do. It must be measurable.

Failure Modes

A failure mode is the manner in which a failure (in the sense of not meeting customer requirements) is observed. The first failure mode defines the boundary of the latitude of a response and may, if the response continues to diverge from what customers require, be superseded by other failure modes, creating a layer effect of failures.

Latitude

This is the range of the response that meets customer requirements. It represents the boundaries of the response that the customer can accept. Outside these boundaries the customer is dissatisfied and this represents a failure in the response.

Uses of the Function Diagram

Depict narrow latitude
Define critical parameters, their nominals and ranges
Define noise factors, their nominals and ranges
Understand how to measure an effective response
Assess sensitivity between specific critical parameters and the response
Assess sensitivity between noise factors and the response
Where a critical parameter is not sensitive assess an opportunity to relax tolerances, reduce cost
Where a critical parameter is highly sensitive to the response assess the capability to manufacture and maintain in control
Define the test methodology based on a full understanding of the functional diagram

CRITICAL PARAMETER MANAGEMENT

Critical Parameter Development

Identification of critical parameters
Design of experimental hardware
Manufacturability assessment
Optimization of critical parameters
Performance test
Redefinition of critical parameters
Design intent established

Critical Parameter Implementation

Implementation of critical parameters on drawings
Comparison of drawings with design intent
Identification of shortfalls and risk
Modification of design intent to meet latitude requirements

Critical Parameter Audit

Critical parameters measured on hardware
Shortfalls and risks identified
Modify hardware of design intent

FAILURE MODES AND EFFECTS ANALYSIS (FMEA)

A process for predicting where failures may occur and devising actions to eliminate or reduce the effect.

Steps in the Process

Define the part and its function(s)
For each function define all potential failure modes
Define the potential effects for each failure mode
Assign a 'severity' score
Define the possible causes of failure
Assign an 'occurrence' score
Define the means for detecting the failure
Assign a 'detection' score
Calculate the risk priority number (RPN):
RPN = severity × occurrence × detection
Prioritize according to the highest RPNs
Assign actions

PROBLEM SOLVING

Identify and select the problem
Analyse the problem
Generate potential solutions
Select and plan a solution
Implement the solution
Evaluate the effectiveness of the solution

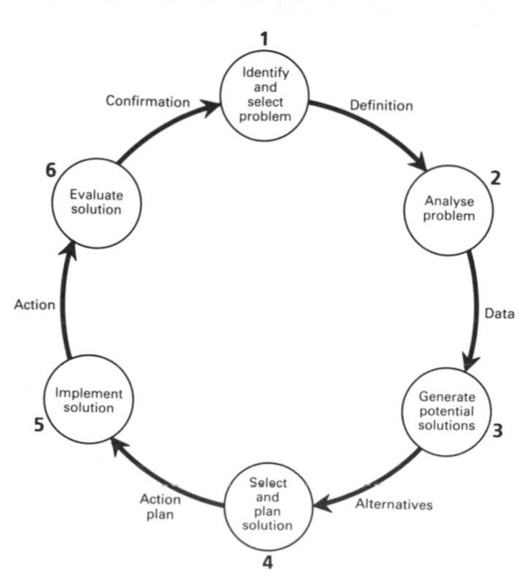

BRAINSTORMING

Rules:

No criticism of ideas
No evaluation of ideas
Take one idea at a time
Write everything down on a flip chart or wall board
Record the words of the contributor faithfully
Work quickly

FISHBONE DIAGRAMS

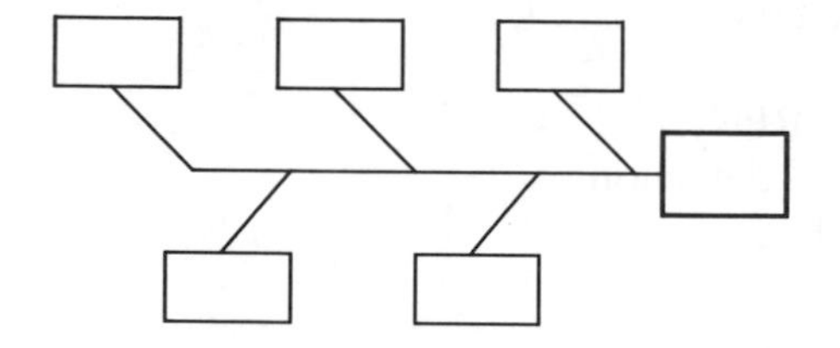

PARETO DIAGRAMS

Vertical bars depicting the values of things that are arranged in descending order

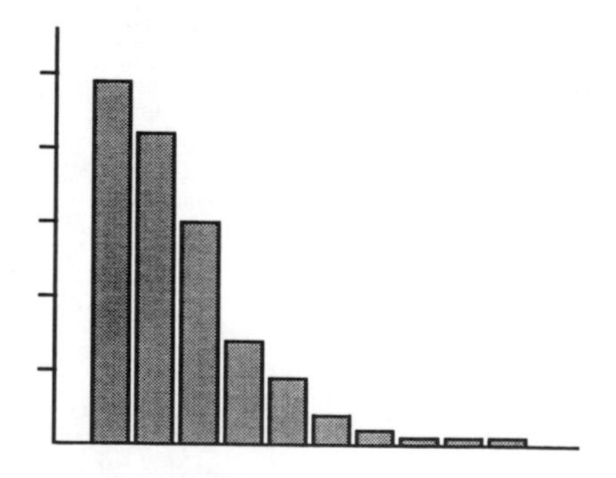

HISTOGRAMS

Charts showing the distribution of values and enabling the evaluation of the amount of variability

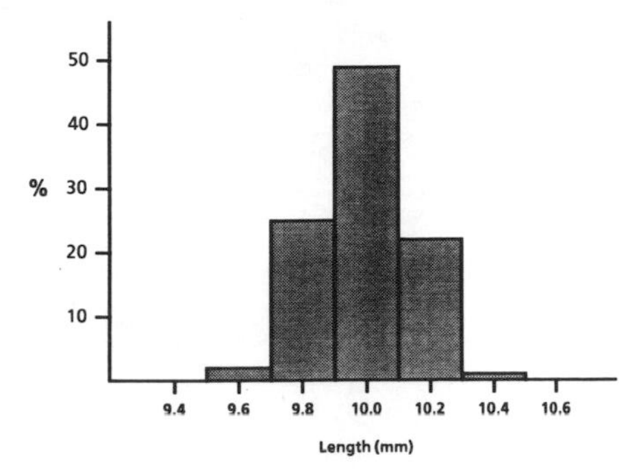

CONTROL CHARTS

A chart that depicts the variation of a particular parameter over time

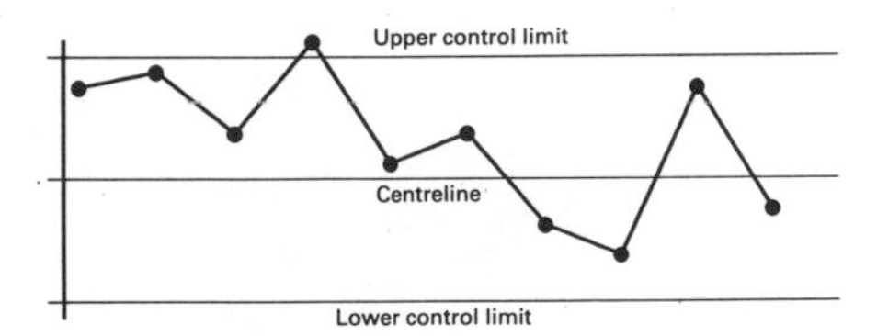

FORCE FIELD DIAGRAM

A diagram showing factors driving a change opposed to factors resisting a change

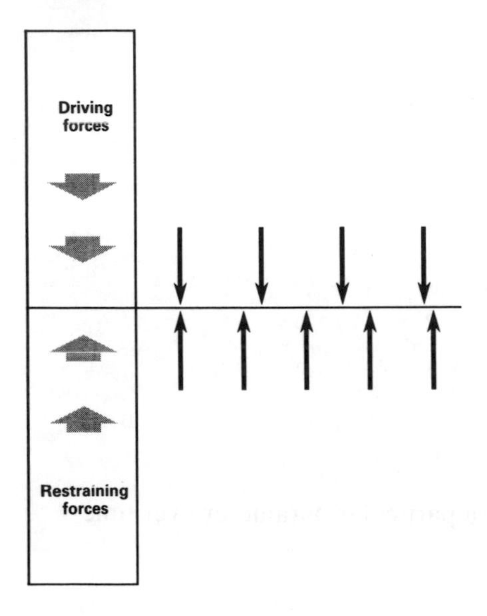

GLOSSARY OF TERMS

Architecture The broad physical interrelationship between parts of a system.

Benchmarking A process of comparison of different techniques, products, services to establish a hierarchy of alternatives.

Brainstorming The uninhibited generation of ideas using group therapy. Individuals in a group environment help each other to stimulate ideas.

Computer aided design (CAD) The use of a computer to develop a design file and detail drawings for parts manufacture.

Control chart A chart showing the variation of a parameter over time.

Control parameters Those parameters that influence the functional aspects of a design. They are usually varied during experimental work.

Critical parameter A design parameter that has a major influence on function and is under the control of the designer. The influence on the function is defined as major when there is a high sensitivity to the response of that function.

Critical parameter audit The checking of hardware for design intent critical parameters.

Critical parameter development The identification and optimization of critical parameters.

Critical parameter implementation The embodiment of critical parameters into a design.

Critical parameter management The process developing, implementing into the design and auditing the critical parameters of a system.

Design for assembly A process to enable a design to be assembled in the easiest and most cost effective way.

Design quality The full and complete set of engineering activities that must be carried out to enable a product to be manufactured that meets customer's requirements.

Failure mode An event in the response of a function which is unacceptable to the customer.

Failure modes and effects analysis (FMEA) A tool that identifies potential failures of a design and enables preventative actions to be put in place.

Fishbone diagram A diagram linking elemental parts to show their influence on the final effect.

Flow diagram A diagram showing the flow of a sequence of events.

Force field diagram A diagram showing all forces influencing a situation. It depicts those forces aiding the situation and those that are in opposition.

Functional analysis system technique (FAST) A diagram and process that represents the functional relationships within a system, enabling an understanding of dependencies between functions and the analysis of cost and reliability for the system.

Hard tooling The tooling that will be used to make the final production parts in large quantities. Hard tooling changes are normally extremely costly and time consuming.

Histogram A bar diagram showing the distribution of a variable.

Latitude The range of responses that are acceptable to the customer as a functional output. These are bounded by failure modes.

Monte Carlo analysis A statistical study of tolerances of parts to predict the final tolerance of the full assembly. Simulates the building of a number of assemblies to reflect the real manufacturing situation.

NASTRAN A stress analysis computer program.

Noise factor A parameter that influences function which is outside the control of the designer.

On-line quality control The practice of checking parts on the production line for their quality level.

Pareto diagram A bar chart which has bars ranked in order of importance.

Production intent design A design that reflects as closely as possible that design which will be in the production product. For example, a production design may have a moulded part. The equivalent part in the production intent design may not be moulded but the design will have as many attributes of the moulding (radii, draft angles, intricate profiles) as possible.

QCD The measurements of quality (Q), cost (C) and time-to-market or delivery (D) of a product under development.

Quality function deployment (QFD) A tool for bringing the 'voice of the customer' to the design and manufacturing process.

Risk priority number The product of the levels of severity, occurrence and detection in the FMEA process.

Robotics The use of automated handling machines (robots) for assembly of parts.

Soft tooling Tooling prepared for a limited number of parts in the pre-production or early stages of production. Soft tooling is usually easier and quicker to create but can only support a small number of production parts. It is often used to develop the design of the hard tooling.

Statistical process control The control of manufacturing processes to ensure specification levels are met.

Supplier integrated computer aided manufacturing (SICAM) The use of a computer to help with the manufacturing of parts from a supplier.

Technology An engineering or scientific method chosen to deliver a specific function. For example, the jet engine and turboprop engine are two technologies for providing thrust for an aircraft. New technologies usually result from invention and may carry higher technical risk.

Tolerance (design tolerance) The range over which a design parameter may vary, that is that which may be tolerated by the design.

Tooling A device to enable the consistent manufacture of a number of identical parts.

Value engineering A process during the design of a product that enables the value to be increased. For example, value may be increased by delivering the same output for less cost, with fewer parts or with higher reliability.

Variance (manufacturing variance) The range over which a manufacturing output varies. Usually this must be kept within the range for design tolerance.

REFERENCES AND BIBLIOGRAPHY

Amsden, Robert T., Howard E. Butler and Davida M. Amsden (1989) *SPC Simplified*, Kraus International Publications, White Plains, New York.

Barker, Thomas B. (1990) *Engineering Quality by Design*, Marcel Dekker, New York.

Bebb, H. Barry (1988) *Design Engineering, The Missing Link in US Competitiveness*, National Academy of Engineering.

Bossert, James L. (1991) *Quality Function Deployment*, Marcel Dekker Inc., New York.

De Bono, Edward (1988) *The Use of Lateral Thinking*, Pelican Books, London.

French, Michael J. (1985) *Conceptual Design for Engineers*, Heinemann Educational, London.

French, Michael J. (1988) *Invention and Evolution*, Cambridge University Press, Cambridge.

Hartley, John and John Mortimer (1991) *Simultaneous Engineering*, Industrial Newsletters Ltd.

Hollins, Bill and Stuart Pugh (1990) *Successful Product Design*, Butterworth, London.

Imai, Masaaki (1986) *Kaisen*, McGraw-Hill, New York.

Kume, Hitoshe (1985) *Statistical Methods for Quality Improvement*, The Association for Overseas Technical Scholarships, Tokyo, Japan.

Middendorf, William H. (1990) *Design of Devices and Systems*, Marcel Dekker, New York.

Miles, Laurence D. (1972) *Techniques for Value Analysis and Engineering*, McGraw-Hill, New York.

Mizuno, Shigeru (1988) *Management for Quality Improvement*, Productivity Press, Cambridge, Mass.

Pugh, Stuart (1991) *Total Design*, Addison-Wesley, Wokingham.

INDEX